SpringerBriefs in Electrical and Computer Engineering

Signal Processing

Series editors

Woon-Seng Gan, Singapore, Singapore
C.-C. Jay Kuo, Los Angeles, USA
Thomas Fang Zheng, Beijing, China
Mauro Barni, Siena, Italy

More information about this series at http://www.springer.com/series/11560

Yong Xiang · Dezhong Peng
Zuyuan Yang

Blind Source Separation

Dependent Component Analysis

 Springer

Yong Xiang
School of Information Technology
Deakin University
Melbourne, Victoria
Australia

Zuyuan Yang
Faculty of Automation
Guangdong University of Technology
Guangzhou, Guangdong
China

Dezhong Peng
College of Computer Science
Sichuan University
Chengdu, Sichuan
China

ISSN 2196-4076   ISSN 2196-4084   (electronic)
ISBN 978-981-287-226-5   ISBN 978-981-287-227-2   (eBook)
DOI 10.1007/978-981-287-227-2

Library of Congress Control Number: 2014940320

Springer Singapore Heidelberg New York Dordrecht London

Printed on acid-free paper

Springer is part of Springer Science+Business Media (www.springer.com)

To my beloved Shan, Angie, and Daniel

Yong

To my beloved family

Dezhong

To my parents

Zuyuan

Preface

"Blind" source separation (BSS) aims to recover unknown source signals from their measurable mixtures without any information about the mixing system. It is a fundamental and challenging problem arising from a wide variety of applications such as wireless communication, medical signal analysis, passive listening, and video surveillance. In traditional BSS methods, a fundamental assumption is that the source signals are independent or at least mutually uncorrelated. This assumption is unfortunately too restrictive to be met in some emerging important applications. For example, in a densely deployed wireless sensor network for microclimatic condition monitoring, the density of sensors may be very high, e.g., more than tens of sensors per square meter. Thus, signals from adjacent sensors are unavoidably cross-correlated and their cross-correlations are unknown. Similarly, in a real-time wireless video surveillance system for anti-terrorism monitoring, images captured by cameras are often mutually correlated. The scenario of mutually correlated signals can also be found in multiple-input multiple-output wireless relay systems, where the signals received and sent by the relay nodes are spatially correlated.

In this book, we provide readers a complete and self-contained set of knowledge about dependent source separation, including the latest development in this field. In Chap. 1, we present an overview of blind source separation. Three promising blind separation techniques that can tackle mutually correlated sources are presented and discussed in Chaps. 2–4, which focus on nonnegativity-based methods, time-frequency analysis-based methods, and precoding-based methods, respectively. Finally, we outline the possible future work in Chap. 5.

Melbourne, Australia, July 2014   Yong Xiang
Chengdu, China   Dezhong Peng
Guangzhou, China   Zuyuan Yang

Acknowledgments

This work was supported in part by the Australian Research Council under grants DP0773446, DP110102076, and LP120100239, the National Basic Research Program of China (973 Program) under grant 2011CB302201, the Program for New Century Excellent Talents in University of China under grants NCET-12-0384 and NCET-13-0740, the National Natural Science Foundation of China under grants 61333013 and 61104053, the Natural Science Foundation of Guangdong Province under Grant S2011030002886, the Specialized Research Fund for the Doctoral Program of Higher Education under Grant 20120181110053, and the Youth Science and Technology Foundation of Sichuan Province under Grant 2013JQ0002.

The authors thank the editors of this series, Prof. Woon-Seng Gan and Prof. C.-C. Jay Kuo, and the Springer team for their constructive guidance and kind assistance.

Acknowledgements

This work was supported in part by the National Natural Science Foundation of China ...

Contents

Chapter 1
Introduction

Abstract Blind source separation (BSS) aims to recover unobserved source signals from their observed mixtures without any information of the mixing system. It is a fundamental problem in signal and image processing. In this chapter, we first introduce the background of BSS, including its history and potential applications. Then, we give a brief overview of the traditional BSS methods for separating independent or uncorrelated source signals. After that, the BSS problem with mutually correlated sources are discussed, together with several mainstream BSS schemes and the corresponding algorithms.

Keywords Blind source separation · Mutually correlated sources · Dependent component analysis

1.1 Background of Blind Source Separation

Generally speaking, blind source separation originates from the well-known cocktail party problem, where a number of people are talking simultaneously in a room and a listener is trying to follow one of the discussions [1]. Mathematically, at a cocktail party (shown in Fig. 1.1), there are m microphones that record or observe r partygoers or speakers at n time increments, where a given microphone is not placed to a given speaker's mouth and is not shielded from the other speakers [2]. Hence, the observed conversations consist of mixtures of true conversations. The problem is to separate or recover the original conversations from the recorded mixed conversations. The human brain can handle this sort of auditory source separation problem to some extent, and BSS tries to solve it like our brain by using a digital signal processing method.

Blind source separation has a wide range of applications in different areas. It is found that many problems can be explained in the framework of BSS, including communication signal processing, biomedical signal identification, remote sensing image interpretation, and so on. In a multiple-input multiple-output (MIMO) wireless communication system, the source signals are transmitted to the receiving end through a MIMO wireless channel system. If multiple users are transmitting signals simultaneously, the signals observed at the receiving antennas will be the mixtures

© The Author(s) 2015
Y. Xiang et al., *Blind Source Separation*,
SpringerBriefs in Signal Processing, DOI 10.1007/978-981-287-227-2_1

Fig. 1.1 Illustration of cocktail party problem as shown in [2]

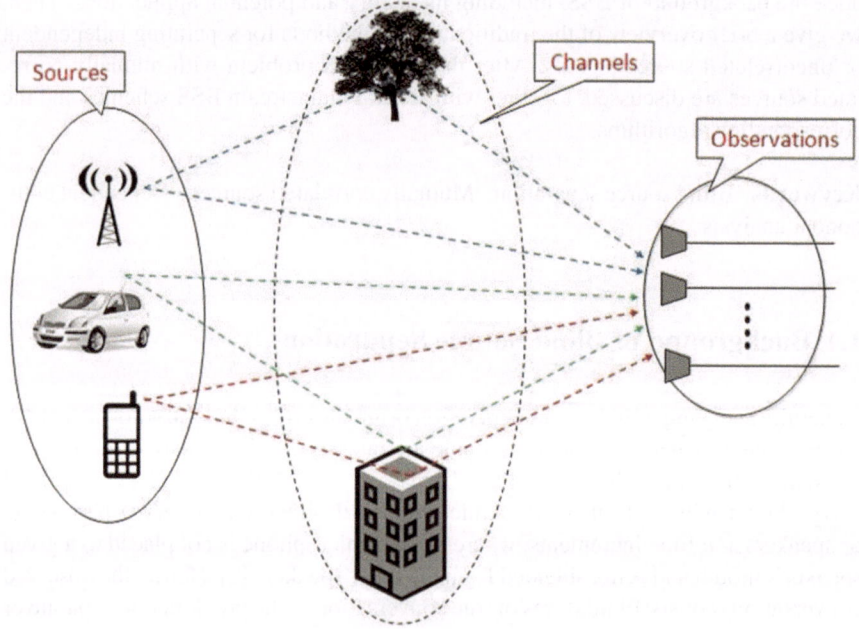

Fig. 1.2 Illustration of a MIMO wireless communication system

of the source signals (see Fig. 1.2) [3, 4]. How to recover the source signals from the received mixtures is a typical BSS problem.

In electroencephalography (EEG) or magnetoencephalography (MEG) signal processing, to explore the activity of the brain, one often puts high density array sensors on the scalp to collect data (see Fig. 1.3) [5, 6]. Due to the resolution restriction of the sensors, the collected EEG or MEG signals are often mixtures of the dipolar

Fig. 1.3 Illustration of EEG
or MEG signal measurement
as shown in [5]

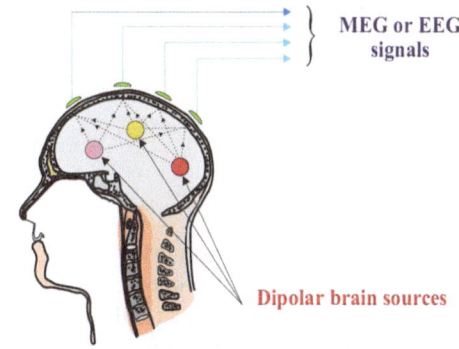

Fig. 1.4 Illustration of
remote sensing imaging as
shown in [7]

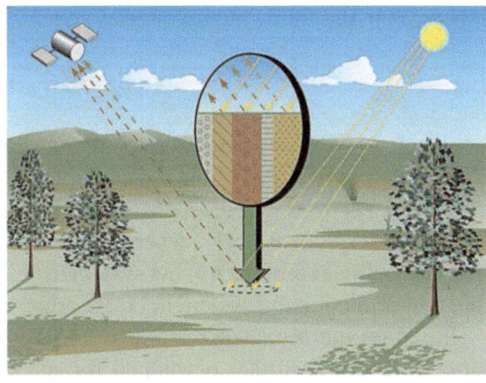

source signals. Then, given the collected EEG or MEG data, how to determine the
source signals of the brain becomes a BSS problem.

In remote sensing image processing, to obtain the grand covers in an unknown
area, a useful method is to scan the area using hyper-sensors with hundreds of spectral
bands located in the airplanes or satellites (see Fig. 1.4) [7]. In the corresponding
hyperspectral imagery, most pixels contain mixed information of more than one
distinct substances or sources, due to the following reasons: a) the spatial resolution
of a sensor is restricted and b) distinct materials are combined into a homogeneous
mixture [7]. Then, the collected spectra are often mixtures, which are mixed by
spectra of different grand covers (or sources) with their proportions (or coefficients).
Obviously, it forms a BSS problem, where one needs to obtain the source spectra
from the observed mixtures without knowing the mixing coefficients.

Regarding the mathematical or physical models for the mixing processes of the
unknown source signals, they may have different styles, depending on the specific
applications. Among those models which can reflect the mixing processes, the instantaneous linear model has the simplest form. Algebraically, most linear BSS models
can be expressed as the following specific matrix factorization [5]:

$$\mathbf{Y} = \mathbf{AX} + \mathbf{W} \tag{1.1}$$

Fig. 1.5 Block diagram of
the mixing system in (1.2)

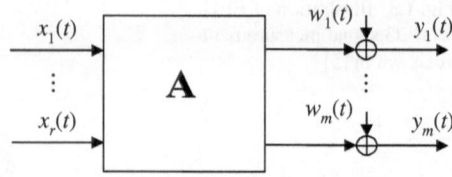

where $\mathbf{Y} \in \mathbb{R}^{m \times n}$ is the observed mixture matrix, $\mathbf{A} \in \mathbb{R}^{m \times r}$ is the mixing matrix, $\mathbf{X} \in \mathbb{R}^{r \times n}$ denotes the source signal matrix, $\mathbf{W} \in \mathbb{R}^{m \times n}$ denotes the noise signal matrix, and m, r and n stand for the number of observations, the number of sources, and the number of samples, respectively. For the BSS problem encountered in some practical applications such as wireless communication, the mixing system is often expressed in the following time series style:

$$\mathbf{y}(t) = \mathbf{A}\mathbf{x}(t) + \mathbf{w}(t) \tag{1.2}$$

where $\mathbf{y}(t) \in \mathbb{R}^{m \times 1}$ is the observed mixed signal vector, $\mathbf{x}(t) \in \mathbb{R}^{r \times 1}$ is the source signal vector, $\mathbf{w}(t) \in \mathbb{R}^{m \times 1}$ is the noise signal vector, and t denotes the time index. Let $\mathbf{x}(t) = [x_1(t), x_2(t), \ldots, x_r(t)]^T$, $\mathbf{y}(t) = [y_1(t), y_2(t), \ldots, y_m(t)]^T$ and $\mathbf{w}(t) = [w_1(t), w_2(t), \ldots, w_m(t)]^T$, where the superscript ^{T} stands for transpose. Assume that the coefficients of the mixing system in (1.2) (or (1.1)) are real valued. The block diagram of the mixing system in (1.2) is shown in Fig. 1.5.

Based on the instantaneous linear mixing model in (1.2), the following convolutional model is further developed for speech and digital communication signal processing [8–11]:

$$\mathbf{y}(t) = \sum_{\tau=-\infty}^{+\infty} \mathbf{A}(\tau)\mathbf{x}(t - \tau) + \mathbf{w}(t) \tag{1.3}$$

where τ denotes the time lag. The convolutional model well reflects the effects of multipath propagation. There also exists the following bridge model [12]:

$$\mathbf{y}(t) = \sum_{\tau=-\infty}^{+\infty} \mathbf{A}\mathbf{x}(t - \tau) + \mathbf{w}(t). \tag{1.4}$$

This model connects the instantaneous model and the convolutional model, i.e., the mixing system is time-invariant and the sources are mixed each other, together with their delayed counterparts.

In addition to the linear mixing models, the nonlinear mixing models have also been proposed for BSS. A representative nonlinear mixing model is the post-nonlinear mixing model [13, 14] shown below:

$$\mathbf{y}(t) = F(\mathbf{A}\mathbf{x}(t)) + \mathbf{w}(t) \tag{1.5}$$

where $F(\cdot)$ denotes a nonlinear function. This model has the nice property that it can be characterized by few parameters. It is a flexible generalization of the standard linear mixing model and can accurately imitate many different nonlinearities.

In this book, we limit our attention to the simplest instantaneous linear mixing model in (1.1) or (1.2). Sometimes, for the sake of simplicity, the noiseless counterparts of (1.1) and (1.2) are also used, which are

$$\mathbf{Y} = \mathbf{AX} \tag{1.6}$$

and

$$\mathbf{y}(t) = \mathbf{Ax}(t) \tag{1.7}$$

respectively.

1.2 Statistics Based Methods for Blind Source Separation

1.2.1 Higher-Order Statistics Based Methods

One classical scheme aiming to solve $\mathbf{x}(t)$ in (1.7) is the so-called independent component analysis (ICA). ICA attempts to decompose a multivariate signal into independent non-Gaussian signals [15]. It gives good results under the assumptions that the source signals are independent of each other and the distribution of the values in each source signal are non-Gaussian. From the viewpoint of implementation, ICA finds the independent components (also called factors, latent variables or sources) by maximizing the statistical independence of the estimated components. Regarding the optimization of independence, there are several versions based on different measures, where two popular ones are minimization of mutual information and maximization of non-Gaussianity. In general, different independence measures give different forms of ICA methods and there exist a series of ICA-based BSS methods. A representative ICA-based BSS method is the fixed-point ICA in [16]. In this method, Comon's information theoretic approach and the projection pursuit approach are combined to obtain independent components, and a family of new contrast (objective) functions for ICA are constructed. Based on these contrast functions, simple fixed-point algorithms are derived, which can optimize the contrast functions very quickly and reliably. The fast ICA-based separation model is as follows [16]:

$$\text{Maxmize}: \ \sum_{i=1}^{r} [E\{G(\mathbf{b}_i^T \mathbf{y}(t))\} - E\{G(v)\}]^2 \tag{1.8}$$

$$\text{s.t. } E\{(\mathbf{b}_j^T \mathbf{y}(t))(\mathbf{b}_k^T \mathbf{y}(t))\} = \delta_{jk}$$

where $\mathbf{b}_i$ denotes the ith column of the separation matrix $\mathbf{B}$, v stands for a Gaussian random variable with zero mean and unit variance, $E\{\cdot\}$ is the mathematical expectation operator, δ_{jk} is the Dirac function which equals one at $j = k$ and zero at $j \neq k$, and $G(\cdot)$ is a nonlinear function, e.g., $G(x) = x^4$. Clearly, ICA-based methods exploit the higher-order statistics (HOS) of the source signals.

The constant modulus (CM) property of signals can also be employed to perform BSS [17] via a multistage approach. This approach first uses a constant modulus algorithm (CMA) to extract one source signal, and then subtracts a replica of the extracted signal from the mixtures. From the result of the subtraction, the CMA can extract another source signal. This process is repeated until all source signals are extracted. In [18], a one-stage CMA-based method is proposed such that all source signals can be estimated at the same iteration. This one-stage method follows the concept of the multi-stage counterpart but modifies the latter by that the cascaded extraction process only occurs within each iteration. Clearly, the error propagation problem still remains in the one-stage method as the separation vectors are updated successively within each iteration. In [19], a separation model based on the constant modulus criterion and the mutual information of the estimated signals is proposed as follows:

$$\text{Minimize}: \quad E\left\{ \sum_{i=1}^{r} \left(\mathbf{e}_i^T \mathbf{B}^T \mathbf{y}(t)\mathbf{y}(t)^T \mathbf{B}\mathbf{e}_i - \gamma \right)^2 \right\}$$

$$+\beta \left\{ \sum_{i=1}^{r} \log \left(\mathbf{e}_i^T \mathbf{B}^T \mathbf{R_{yy}}\mathbf{B}\mathbf{e}_i \right) - \log \left(\det(\mathbf{B}^T \mathbf{R_{yy}}\mathbf{B}) \right) \right\} \quad (1.9)$$

where

$$\mathbf{R_{yy}} \triangleq E\left(\mathbf{y}(t)\mathbf{y}(t)^T \right) \quad (1.10)$$

is the autocorrelation matrix of $\mathbf{y}(t)$, $\mathbf{e}_i$ is an $r \times 1$ column vector whose ith element is one and the rest elements are zero, γ is *a priori* constant dispersion, β is a positive real number that trades off the amplitude term and the mutual information term, and $\log(\cdot)$ and $\det(\cdot)$ are respectively the logarithm and determinant functions. The algorithm derived from this cost function directly updates the separation matrix rather than successively updates the separation vectors. Consequently, it separates all source signals simultaneously and also eliminates the propagation of estimation errors. Clearly, the CMA-based methods also utilize the HOS of the mixtures.

1.2.2 Second-Order Statistics Based Methods

It is interesting to note that the BSS methods that utilize the HOS of the mixtures usually require the source signals to be white. While white signals are often encountered in practical applications, colored signals also widely exist. In fact, information bearing signals are often colored. The color of a signal may be inherent in the original message, or be built into the signal by the system designer. By exploiting the temporal correlation of signals, alternative methods can be developed based on second-order statistics (SOS), instead of HOS. The HOS-based methods usually need a large number of data samples to obtain satisfactory performance. Thus, in applications where fewer data samples are available, the SOS-based methods are preferable [4, 20–22].

Among the existing SOS-based BSS methods, the approximate joint diagonalization algorithms play crucial roles. They are generally divided into two categories: orthogonal joint diagonalization (OJD) and nonorthogonal joint diagonalization (NJD). The OJD algorithms restrict the diagonalizer to be orthogonal and are applicable to BSS when the observations are prewhitened. However, because of some disadvantages of the prewhitening phase in BSS, the NJD algorithms have received increasing attention in recent years. Most NJD algorithms are based on the following criterion of minimizing diagonalization error [23]:

$$\text{Minimize}: \ \sum_{k=1}^{K} \text{off}(\mathbf{B}^T \mathbf{R_{yy}}(\tau_k)\mathbf{B}) \tag{1.11}$$

where $\text{off}(\mathbf{A}) = \sum_{i \neq j} a_{ij}^2$, $\mathbf{R_{yy}}(\tau_k)$ is the auto-correlation matrix of $\mathbf{y}(t)$ with time lag τ_k, and K denotes the number of time lags. Generally, two important things are involved in the NJD methods: (i) impose proper constraints to obtain reasonable solutions and (ii) design an efficient algorithm to solve the corresponding optimization model. Regarding the constraints, since a good diagonalizer should have a small condition number when it minimizes the diagonalization error [23], the constraints can be uniformly expressed in terms of condition number, i.e., minimizing the condition number of the diagonalizers. Then, the approximate joint diagonalization problem can be treated as a multi-objective optimization problem. To solve this multi-objective optimization problem, a simple algorithm is proposed in [23]. It achieves the minimum diagonalization error and the corresponding diagonalizer has a small condition number. Furthermore, it imposes few restrictions on the target set of matrices to be diagonalized, making it applicable to different applications.

Another SOS-based method is the one based on temporal predictability, where the source signals and their mixtures are assumed to have distinct temporal predictability [24]. Like the traditional SOS-based methods, the temporal predictability method only assumes that the sources are uncorrelated. It does not need to estimate the probability density functions and can separate super-Gaussian signals and sub-Gaussian

signals simultaneously. This method is based on the conjecture that the temporal predictability of any signal mixture is less than (or equal to) that of any of its component
source signals. The conjecture is improved in an equivalent concept of covariance
rate in [25]. It is proved that the covariance rate of a mixture signal is between the
maximal and minimal covariance rates of the sources. In [26], it is further shown
that the sources are separable by the temporal predictability method if and only if
they have different temporal structures (i.e., autocorrelations), which indicates the
applicability and limitations of the temporal predictability method. Meanwhile, the
following separation model is proposed for estimating the separation vectors $\mathbf{b}_i$,
$i = 1, 2, \ldots, r$ that form the separation matrix $\mathbf{B}$ [26]:

$$\text{Maximize} : \frac{\mathbf{b}_i^T \mathbf{C}_{yy}^{\lambda_L} \mathbf{b}_i}{\mathbf{b}_i^T \mathbf{C}_{yy}^{\lambda_S} \mathbf{b}_i} \qquad (1.12)$$

where $\lambda_L = 2^{-\frac{1}{h_L}}, \lambda_S = 2^{-\frac{1}{h_S}}, 0 < h_S \ll h_L$ are parameters, and $\mathbf{C}_{yy}^{\lambda}$ is the
covariance matrix of $\mathbf{y}(t)$ with time lag controlled by λ.

In [27], Abed-Meraim *et al.* show that the second-order cyclic statistics can also
be applied to BSS. Specifically, blind source separation can be achieved using the
cyclic correlation matrices of the mixed signals if there do not exist two distinct
source signals whose cycle frequencies are the same and whose cyclic autocorrelation vectors are linearly dependent. A set of algorithms are proposed in [27] for
blindly recovering the source signals. In [28], the phase redundancy and frequency
redundancy of cyclostationary signals are exploited to tackle the BSS problem in
a complementary way. This method requires a weaker separation condition than
those methods which only exploit the phase diversity or the frequency diversity of
the source signals. The separation criterion is to diagonalize a polynomial matrix
whose coefficient matrices consist of the correlation and cyclic correlation matrices
of multiple measurements.

1.3 Blind Source Separation via Dependent Component Analysis

1.3.1 Scenarios of Mutually Correlated Sources

While the condition that the source signals are mutually independent or at least
uncorrelated is reasonable for many applications, spatially correlated sources also
occur in practice. For example, in recent years, the Internet of Things (IOT) based
on wireless sensor network has attracted a lot of research interest [29–31]. In order
to provide high reliability in face of the failure of individual sensors, and/or facilitate
superior spatial localization of objects of interest, some wireless sensors could be
densely deployed in IOT. As a result, signals from adjacent sensors are unavoidably
cross correlated [32] and their cross correlations are usually unknown. Correlation

Fig. 1.6 Correlated natural images

may also arise between signals from nonadjacent sensors if the sensors sample an environmental parameter that does not change significantly at their locations.

Besides, if each sensor is equipped with a video camera, the IOT becomes a wireless video surveillance system. It is known that images which look irrelevant are often mutually correlated [33–35]. Especially, human face images show particularly strong correlations [36]. These conclusions can be easily verified by using some real world images. For instance, Fig. 1.6 shows four different images, which look irrelevant content-wise. If one computes their correlation coefficient matrix, it yields

$$\mathbf{C}_1 = \begin{bmatrix} 1.0000 & 0.3461 & 0.4020 & -0.5427 \\ 0.3461 & 1.0000 & 0.7022 & -0.5053 \\ 0.4020 & 0.7022 & 1.0000 & -0.4589 \\ -0.5427 & -0.5053 & -0.4589 & 1.0000 \end{bmatrix}.$$

Fig. 1.7 Human face images with strong correlation

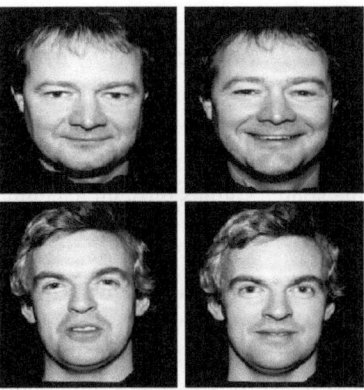

Clearly, these images are mutually correlated as the off-diagonal entries of $\mathbf{C}_1$ are nonzero. Similarly, for the face images in Fig. 1.7, the corresponding correlation coefficient matrix is

$$\mathbf{C}_2 = \begin{bmatrix} 1.0000 & 0.9106 & 0.8265 & 0.8421 \\ 0.9106 & 1.0000 & 0.8237 & 0.8379 \\ 0.8265 & 0.8237 & 1.0000 & 0.8451 \\ 0.8421 & 0.8379 & 0.8451 & 1.0000 \end{bmatrix}.$$

Since the off-diagonal entries of $\mathbf{C}_2$ are very large, these images have strong correlations. More examples of correlated images can be found in fingerprint images [37], X-ray images [38, 39], abundance images [7], and so on.

Mutually correlated signals can also be found in MIMO wireless relay systems [40–43], where the communication between the source and destination nodes is carried out via two stages. In the first stage, the source nodes transmit signals to the relay nodes. In the second stage, the source nodes are silent, and each relay node amplifies the received signal and then sends the amplified signal to the destination node. In these relay systems, the instantaneous channel state information (CSI) of the relay-destination link is essential to the optimization of the relay amplifying power [40–43]. However, since the instantaneous CSI is unknown in practical relay communication systems, it has to be estimated at the destination node [44, 45]. This problem can be cast as a BSS problem but the signals received and sent by the relay nodes are spatially correlated.

In the case of mutually correlated sources mentioned above, the conventional BSS methods exploiting the higher-order or second-order statistical properties of the sources will fail. To develop BSS methods for mutually correlated sources, one must analyze and exploit other properties of the source signals. We call this dependent component analysis (DCA), in contrast to the well-known ICA technique used for the blind separation of mutually independent (or uncorrelated) sources.

1.3.2 Dependent Component Analysis Based Methods

The current state-of-the-art works for BSS via DCA employ some special features of the source signals, such as nonnegativity and sparsity, or build some features into the source signals through a technique called precoding.

1.3.2.1 Exploiting Nonnegativity

Nonnegative signals exist in various applications such as biomedical data analysis and image processing. One popular BSS method exploiting source nonnegativity is called convex analysis of mixtures of nonnegative sources by linear programming (CAMNS-LP) [38]. In addition to nonnegativity, this method also exploits a special source feature named *local dominance*, which means that for each source there is at least one time instant at which the source dominates. This assumption is considered to be realistic for many real-world signals such as human portraits and some multichannel biomedical images. Under the *local dominance* assumption and some other standard assumptions, a BSS criterion is established through convex analysis. According to this criterion, the source signals can be perfectly identified by finding the extreme points of an observation-constructed polyhedral set, and the corresponding separation model is as follows [38]:

$$\text{Maximize}: \ \mathbf{r}^T (\mathbf{C}\alpha + \mathbf{d}) \tag{1.13}$$

$$\text{s.t. } \mathbf{C}\alpha + \mathbf{d} \succeq \mathbf{0}$$

where α is the variable vector, $\mathbf{r}$ is a random vector, $(\mathbf{C}, \mathbf{d})$ is the 2-tuple related only to the observed mixtures $\mathbf{Y}$, and $\succeq$ is component-wise. After an optimal solution α^* is calculated (which means that one extreme point is found), one source is estimated by $\mathbf{C}\alpha^* + \mathbf{d}$. Since both the maximum and the minimum may correspond to the extreme point, the maximization can also be replaced by the minimization. The CAMNS-LP method provides an effective tool for the blind separation of correlated sources but its speed is restricted by the time-consuming computation in solving multiple LP problems.

Recently, a fast project pursuit (PP) based method has been developed in [46]. The PP method first maps the available observation matrix $\mathbf{Y}$ into a super-plane to construct a new mixing model, in which the inaccessible source matrix $\mathbf{X}$ is normalized to be column-sum-to-one and those columns of the normalized source matrix that satisfy the local dominance condition are unit vectors consequently. Then, based on the property of the normalized source matrix, it solves the new mixing model to estimate the mixing matrix $\mathbf{A}$. The mixing matrix is estimated column by column through tracing the projections to the mapped observations in specified directions. Once the mixing matrix is obtained, the source signals can be easily recovered from the observed mixtures. The PP method is much faster than the CAMNS-LP method in [38] because it only needs to solve one LP problem.

Another well-known approach to processing nonnegative signals is nonnegative matrix factorization (NMF). NMF aims to decompose a given nonnegative matrix into two nonnegative factor matrices [47]. Since the mathematical model of NMF is similar to BSS and it does not rely on the statistical information of the decompositions, NMF has great potential to be applied to separating mutually correlated sources [48, 49]. However, generally speaking, NMF does not necessarily generate a desired result, and thus one often needs to add some constraints to the standard NMF in order to solve the BSS problem. Some constrained NMF methods have been developed to perform BSS, such as the minimum volume constrained NMF [50] and the minimum dispersion constrained NMF [51].

1.3.2.2 Exploiting Nonnegativity and Sparsity in Time Domain

Sparsity is another property existing in numerous signals such as images. Exploiting both the nonnegativity and sparsity features of signals, various BSS methods have been developed [52, 53]. For instance, in [52], a sparse NMF based separation model using Euclidean distance is proposed for BSS as:

$$\text{Minimize}: \frac{1}{2}\|\mathbf{Y} - \mathbf{AX}\|_2^2 + \beta\|\mathbf{X}\|_1 \qquad (1.14)$$

$$\text{s.t. } \mathbf{A} \succeq \mathbf{0} \text{ and } \mathbf{X} \succeq \mathbf{0}$$

where $\beta > 0$ and $\|\cdot\|_1$ denotes the L_1-norm.

In order to utilize the sparsity feature of sources, an important issue is to find a proper mathematical measure to gauge the sparseness of the sources [54]. It is known that for a single signal, the sparseness can be gauged by many popular sparseness measures, such as Donoho's measure [55] which is based on the L_0-norm of the signal (i.e., the number of zero elements) and Hoyer's measure [56] which is based on the normalized ratio of the L_1-norm and L_2-norm of the signal (i.e., the ratio of the absolute sum of the elements and the squared root of the quadratic sum of the elements). However, these sparseness measures do not reflect the joint sparseness of multiple sources. In order to describe the joint sparseness of the nonnegative sources, a determinant-based sparseness measure, called D-measure, is developed in [37]. Based on the D-measure, the following separation model is proposed [37]:

$$\text{Maximize}: \det(\mathbf{BYY}^T\mathbf{B}^T) \qquad (1.15)$$

$$\text{s.t. } \mathbf{BY} \succeq \mathbf{0}$$

from which an iterative sparseness maximization approach is derived. In this approach, the BSS problem is cast into row-by-row optimizations with respect to the separation matrix, and the quadratic programming is invoked to optimize each row. This approach needs neither the *local dominance* assumption required by the convex analysis based methods [38, 46], nor the selection of the balance parameters encountered in the constrained NMF algorithms [50, 51].

1.3.2.3 Exploiting Sparsity in Time-Frequency Domain

It should be noted that all of the BSS methods mentioned above usually require the mixing matrix to be determined, i.e., the number of the mixtures is no less than that of the sources. However, sometimes, the former could be less than the latter, i.e., the mixing matrix is underdetermined. Moreover, the source signals may not be nonnegative. In the underdetermined case, recovering the source signals is a very challenging task, even if the mixing matrix is known, since the mixing matrix is not invertible. The source signals could be retrieved if they are sufficiently sparse. Since the representations of signals are normally sparser in time-frequency (TF) domain than in time domain, the TF representations of signals are commonly used in this case. Regarding TF representations, two types of common transforms are [57]: (i) linear TF transforms, e.g., short-time Fourier transform and wavelet transform; and (ii) quadratic TF transforms, e.g., Wigner-Ville transform and Cohen's class of transform. Based on the TF representations of the source signals, the time-frequency analysis (TFA) has been used as an effective tool to develop underdetermined blind source separation (UBSS) methods [58–64].

For the UBSS methods in [61, 62], it is assumed that the source signals are completely sparse in the TF domain, i.e., TF-disjoint. It means that there is at most one active source at any TF point, which is a restrictive constraint and is difficult to meet in practice. In order to overcome this drawback, two subspace-based methods are proposed for TF-nondisjoint sources in [63], which assumes that at any TF point, the number of active sources is less than that of the mixtures and any two column vectors of the mixing matrix are pairwise linearly independent. In [64], Peng and Xiang show that to ensure the subspace-based methods in [63] to work, the condition that the column vectors of the mixing matrix are pairwise linearly independent is not sufficient. Instead, any $m \times m$ submatrix of the mixing matrix should be full column rank, where m is the number of the mixtures. Besides, a UBSS method is developed in [64], which needs a weaker sparsity condition on the source signals, i.e., the number of active sources at any TF point should be less than or equal to the number of the mixtures.

The requirement on the source signals is further relaxed in the method called UBSS with free active sources (UBSS-FAS), which is presented in [59]. Specifically, if the number of the sources is less than twice of that of the mixtures, the sources can be recovered exactly at every auto-term TF point no matter how many active sources there are. Besides, it is interesting to note that if the auto-source points and cross-source points of the source signals do not overlap in the TF plane, the sources mixed by a known underdetermined mixing system can be separated by using TFA, even though the sources are non-sparse [65]. However, estimating the mixing matrix which has more columns than rows usually has to exploit the sparsity property of the sources.

1.3.2.4 Utilizing Precoding

The afore-mentioned methods for the blind separation of dependent sources require the source signals to be nonnegative and/or sparse. However, these conditions may not exist in some applications. For example, most communication signals are neither nonnegative nor sparse. Nevertheless, these communication signals are accessible on the transmission side. To facilitate the blind separation of such sources at the receiving end, an intuition of thinking is to preprocess the source signals before transmission to enhance their spatial diversity. An approach to achieving this is to use precoders in transmitters. For example, a precoding scheme is used in [66] to reduce the cross-correlation of the sources, while a prefiltering approach is employed in [67] to enhance source independence. In [68], a precoding based method is proposed to separate mutually correlated sources in time domain, without resort to the nonnegativity and/or sparsity of the sources. It is shown that by applying properly designed precoders to the correlated source signals, the coded signals have zero cross-correlation at some time lags. Based on this special feature, a closed-form algorithm is derived to separate the coded signals at the receiver and then estimate the source signals.

The order of the precoders used in the method in [68] is $4r - 1$, where r is the number of the sources. Obviously, with the increase of sources, the order of the precoders will rise proportionally, which will increase computation complexity and signal transmission delays. In [69], a Z-domain method is proposed to separate spatially correlated sources, which utilizes the Z-domain structure of the coded signals. The order of the precoders used in this method is only two, regardless of the number of sources. In [70], the order of the precoders is further reduced to one, again independent of the number of sources.

In the reminder of the book, we will investigate the BSS problem via DCA in detail, and present and discuss the latest methods for blind separation of dependent sources. Particularly, in Chap. 2, we introduce the BSS methods that exploit the nonnegativity and/or sparsity properties of the sources in time domain. Chapter 3 presents the methods for underdetermined BSS or UBSS, which employ the sparsity property or other special property of the sources in time-frequency domain. The precoding based BSS methods are shown in Chap. 4, which do not require the sources to be nonnegative or sparse. Chapter 5 briefly outlines possible future work.

References

1. P. Comon, C. Jutten, *Handbook of Blind Source Separation: Independent Component Analysis and Applications, 1ed* (Academic, Oxford, 2010)
2. D.B. Rowe, *Multivariate Bayesian Statistics: Models for Source Separation and Signal Unmixing* (CRC, Boca Raton, 2002)
3. J. Wang, C. Lin, E. Siahaan, B. Chen, H. Chuang, Mixed sound event verification on wireless sensor network for home automation. IEEE Trans. Industr. Inf. **10**(1), 803–812 (2014)
4. S. An, Y. Hua, J.H. Manton, Z. Fang, Group decorrelation enhanced subspace method for identifying FIR MIMO channels driven by unknown uncorrelated colored sources. IEEE Trans. Signal Process. **53**(12), 4429–4441 (2005)
5. A. Cichocki, Blind signal processing methods for analyzing multichannel brain signals. Int. J. Bioelectromagnetism **6**, 1 (2004)
6. P. Xu, Y. Tian, H. Chen, D. Yao, L_p norm iterative sparse solution for EEG source localization. IEEE Trans. Biomed. Eng. **54**(3), 400–409 (2007)
7. N. Keshava, J.F. Mustard, Spectral unmixing. IEEE Signal Process. Mag. **19**(1), 44–57 (2002)
8. H. Sawada, S. Araki, S. Makino, Underdetermined convolutive blind source separation via frequency bin-wise clustering and permutation alignment. IEEE Trans. Audio Speech Lang. Process. **19**(3), 516–527 (2011)
9. Y. Xiang, V.K. Nguyen, N. Gu, Blind equalization of nonirreducible systems using CM criterion. IEEE Trans. Circuits Syst. II Express Briefs **53**(8), 758–762 (2006)
10. Y. Xiang, S. Nahavandi, H. Trinh, H. Zheng, A new second order method for blind signal separation from dynamic mixtures. Comput. Electr. Eng. **30**(5), 347–359 (2004)

11. H.A. Inan, A.T. Erdogan, Convolutive bounded component analysis algorithms for independent and dependent source separation. to appear in IEEE Trans. Neural Networks Learn. Syst. doi:10.1109/TNNLS.2014.2320817
12. P. Bofill, Underdetermined blind separation of delayed sound sources in the frequency domain. Neurocomputing **55**(3–4), 627–641 (2003)
13. Y. Altmann, N. Dobigeon, J.-Y. Tourneret, Nonlinearity detection in hyperspectral images using a polynomial post-nonlinear mixing model. IEEE Trans. Image Proc. **22**(4), 1267–1276 (2013)
14. A. Taleb, C. Jutten, Source separation in post-nonlinear mixtures. IEEE Trans. Signal Proc. **47**(10), 2807–2820 (1999)
15. J.V. Stone, *Independent Component Analysis: A Tutorial Introduction* (MIT, Cambridge, 2004)
16. A. Hyvärinen, Fast and robust fixed-point algorithms for independent component analysis. IEEE Trans. Neural Networks **10**(3), 626–634 (1999)
17. J.J. Shynk, A.V. Keerthi, A. Mathur, Steady-state analysis of the multistage constant modulus array. IEEE Trans. Signal Proc. **44**(4), 948–962 (1996)
18. A. Touzni, I. Fijalkow, M.G. Larimore, J.R. Treichler, A globally convergent approach for blind MIMO adaptive deconvolution. IEEE Trans. Signal Proc. **49**(6), 1166–1178 (2001)
19. Y. Xiang, Blind source separation based on constant modulus criterion and signal mutual information. Comput. Electr. Eng. **34**(5), 416–422 (2008)
20. G. Chabriel, M. Kleinsteuber, E. Moreau, H. Shen, P. Tichavsky, A. Yeredor, Joint matrices decompositions and blind source separation: a survey of methods, identification, and applications. IEEE Signal Process. Mag. **31**(3), 34–43 (2014)
21. Y. Xiang, S.K. Ng, An approach to nonirreducible MIMO FIR channel equalization. IEEE Trans. Circuits Syst. II Express Briefs **56**(6), 494–498 (2009)
22. Y. Xiang, L. Yang, D. Peng, S. Xie, A second-order blind equalization method robust to ill-conditioned SIMO FIR channels. Digit. Signal Proc. **32**, 57–66 (2014)
23. G. Zhou, S. Xie, Z. Yang, J. Zhang, Nonorthogonal approximate joint diagonalization with well-conditioned diagonalizers. IEEE Trans. Neural Networks **20**(11), 1810–1819 (2009)
24. J.V. Stone, Blind source separation using temporal predictability. Neural Comput. **13**(7), 1559–1574 (2001)
25. S. Xie, Z. He, Y. Fu, A note on Stone's conjecture of blind signal separation. Neural Comput. **17**(2), 321–330 (2005)
26. S. Xie, G. Zhou, Z. Yang, Y. Fu, On blind separability based on the temporal predictability method. Neural Comput. **21**(12), 3519–3531 (2009)
27. K. Abed-Meraim, Y. Xiang, J.H. Manton, Y. Hua, Blind source separation using second-order cyclostationary statistics. IEEE Trans. Signal Proc. **49**(4), 694–701 (2001)
28. Y. Xiang, W. Yu, J. Zhang, S. An, Blind source separation based on phase and frequency redundancy of cyclostationary signals. IEICE Trans. Fundam. Electron. Commun. Comput. Sci. **E87**—**A**(12), 3343–3349 (2004)
29. L. Qu, Y. Huang, C. Tang, T. Han, Node design of internet of things based on ZigBee multi-channel. Procedia Eng. **29**, 1516–1520 (2012)
30. F. Mattern, C. Floerkemeier, From the internet of computers to the internet of things. Lect. Notes Comput. Sci. **6462**, 242–259 (2010)
31. L. Atzori, A. Iera, G. Morabito, The internet of things: a survey. Comput. Netw. **54**(15), 2787–2805 (2010)
32. S.S. Pradhan, J. Kusuma, K. Ramchandran, Distributed compression in a dense microsensor network. IEEE Signal Process. Mag. **19**(2), 51–60 (2002)
33. C.-Y. Chang, A.A. Maciejewski, V. Balakrishnan, Fast eigenspace decomposition of correlated images. IEEE Trans. Image Proc. **9**(11), 1937–1949 (2000)
34. K. Saitwal, A.A. Maciejewski, R.G. Roberts, B.A. Draper, Using the low-resolution properties of correlated images to improve the computational efficiency of eigenspace decomposition. IEEE Trans. Image Proc. **15**(8), 2376–2387 (2006)
35. Y. Peng, A. Ganesh, J. Wright, W. Xu, Y. Ma, RASL: Robust alignment by sparse and low-rank decomposition for linearly correlated images, in *Proc. 2010 IEEE Computer Society Conference on Computer Vision and Pattern Recognition*, 2000, art. no. 5540138, pp. 763–770

36. G. Zhou, Z. Yang, S. Xie, J. Yang, Online blind source separation using incremental nonnegative matrix factorization with volume constraint. IEEE Trans. Neural Networks **22**(4), 550–560 (2011)

37. Z. Yang, Y. Xiang, S. Xie, S. Ding, Y. Rong, Nonnegative blind source separation by sparse component analysis based on determinant measure. IEEE Trans. Neural Netw. Learn. Syst. **23**(10), 1601–1610 (2012)

38. T.H. Chan, W.K. Ma, C.Y. Chi, Y. Wang, A convex analysis framework for blind separation of non-negative sources. IEEE Trans. Signal Proc. **56**(10), 5120–5134 (2008)

39. K. Suzuki, R. Engelmann, H. MacMahon, K. Doi, Virtual dual-energy radiography: improved chest radiographs by means of rib suppression based on a massive training artificial neural network (Mtann), Radiology, vol. 238, 2006 http://suzukilab.uchicago.edu/research.htm,

40. X. Tang, Y. Hua, Optimal design of non-regenerative MIMO wireless relays. IEEE Trans. Wireless Commun. **6**(4), 1398–1407 (2007)

41. I. Hammerström, A. Wittneben, Power allocation schemes for amplify-and-forward MIMO-OFDM relay links. IEEE Trans. Wireless Commun. **6**(8), 2798–2802 (2007)

42. A.S. Behbahani, R. Merched, A.M. Eltawil, Optimizations of a MIMO relay network. IEEE Trans. Signal Proc. **56**(10), 5062–5073 (2008)

43. Y. Rong, M.R.A. Khandaker, Y. Xiang, Channel estimation of dual-hop MIMO relay system via parallel factor analysis. IEEE Trans. Wireless Commun. **11**(6), 2224–2233 (2012)

44. P. Lioliou, M. Viberg, M. Coldrey, Performance analysis of relay channel estimation, in *Proceedings of IEEE Asilomar*, CA, USA, Nov, Pacific Grove, 2009, pp. 1533–7

45. F. Gao, T. Cui, A. Nallanathan, On channel estimation and optimal training design for amplify and forward relay networks. IEEE Trans. Wireless Commun. **7**(5), 1907–1916 (2008)

46. Z. Yang, Y. Xiang, Y. Rong, S. Xie, Projection-pursuit-based method for blind separation of nonnegative sources. IEEE Trans. Neural Netw. Learn. Syst. **24**(1), 47–57 (2013)

47. D.D. Lee, H.S. Seung, Learning of the parts of objects by non-negative matrix factorization. Nature **401**(6755), 788–791 (1999)

48. A. Cichocki, R. Zdunek, S. I. Amari, New algorithms for nonnegative matrix factorization in applications to blind source separation, in *Proceedings 2006 IEEE International Conference on Acoustics, Speech and Signal Processing*, 2006, pp. 5479–5482

49. B. Gao, W.L. Woo, B.W.-K. Ling, Machine learning source separation using maximum a posteriori nonnegative matrix factorization. IEEE Trans. Cybern. **44**(7), 1169–1179 (2014)

50. L. Miao, H. Qi, Endmember extraction from highly mixed data using minimum volume constrained nonnegative matrix factorization. IEEE Geosci. Remote Sens. **45**(3), 765–777 (2007)

51. A. Huck, M. Guillaume, J. Blanc-Talon, Minimum dispersion constrained nonnegative matrix factorization to unmix hyperspectral data. IEEE Trans. Geosci. Remote Sens. **48**(6), 2590–2612 (2010)

52. V.P. Pauca, J. Piper, R.J. Plemmons, Nonnegative matrix factorization for spectral data analysis. Linear Algebra Appl. **416**(1), 29–47 (2006)

53. Z. Yang, G. Zhou, S. Xie, S. Ding, J. Yang, J. Zhang, Blind spectral unmixing based on sparse nonnegative matrix factorization. IEEE Trans. Image Proc. **20**(4), 1112–1125 (2011)

54. N. Hurley, S. Rickard, Comparing measures of sparsity. IEEE Trans. Inf. Theory **55**(10), 4723–4741 (2009)

55. D.L. Donoho, M. Elad, Optimally sparse representation in general (nonorthogonal) dictionaries via ℓ1 minimization. PNAS **100**(5), 2197–2202 (2003)

56. P.O. Hoyer, Non-negative matrix factorization with sparseness constraints. J. Mach. Learn. Res. **5**, 1457–1469 (2004)

57. S. Qian, D. Chen, Joint time-frequency analysis. IEEE Signal Process. Mag. **16**(2), 52–67 (1999)

58. L. Cirillo, A. Zoubir, M. Amin, Blind source separation in the time-frequency domain based on multiple hypothesis testing. IEEE Trans. Signal Process. **56**(6), 2267–2279 (2008)

59. S. Xie, L. Yang, J. Yang, G. Zhou, Y. Xiang, Time-frequency approach to underdetermined blind source separation. IEEE Trans. Neural Netw. Learn. Syst. **23**(2), 306–316 (2012)

60. A.L.F. de Almeida, G. Favier, Double Khatri-Rao space-time-frequency coding using semi-blind PARAFAC based receiver. IEEE Signal Process. Lett. **20**(5), 471–474 (2013)
61. N. Linh-Trung, A. Belouchrani, K. Abed-Meraim, B. Boashash, Separating more sources than sensors using time-frequency distributions. EURASIP J. Appl. Signal Process. **17**, 2828–2847 (2005)
62. Y. Luo, W. Wang, J.A. Chambers, S. Lambotharan, I. Proudler, Exploitation of source non-stationarity in underdetermined blind source separation with advanced clustering techniques. IEEE Trans. Signal Process. **54**(6), 2198–2212 (2006)
63. A. Aissa-El-Bey, N. Linh-Trung, K. Abed-Meraim, A. Belouchrani, Y. Grenier, Underdetermined blind separation of nondisjoint sources in the time-frequency domain. IEEE Trans. Signal Process. **55**(3), 897–907 (2007)
64. D. Peng, Y. Xiang, Underdetermined blind source separation based on relaxed sparsity condition of sources. IEEE Trans. Signal Process. **57**(2), 809–814 (2009)
65. D. Peng, Y. Xiang, Underdetermined blind separation of non-sparse sources using spatial time-frequency distributions. Digit. Signal Process. **20**(2), 581–596 (2010)
66. S. Zhou, G.B. Giannakis, Optimal transmitter eigen-beamforming and space-time block coding based on channel correlations. IEEE Trans. Inf. Theory **49**(7), 1673–1690 (2003)
67. A.-J. van der Veen, Joint diagonalization via subspace fitting techniques. Proc. Int. Conf. Acoust. Speech Signal Process. **5**, 2773–2776 (2001)
68. Y. Xiang, S.K. Ng, V.K. Nguyen, Blind separation of mutually correlated sources using pre-coders. IEEE Trans. Neural Networks **21**(1), 82–90 (2010)
69. Y. Xiang, D. Peng, Y. Xiang, S. Guo, Novel Z-domain precoding method for blind separation of spatially correlated signals. IEEE Trans. Neural Netw. Learn. Syst. **24**(1), 94–105 (2013)
70. Y. Xiang, D. Peng, A. Kouzani, Separating spatially correlated signals using first-order pre-coders, in *Proceeding of The 9th IEEE Conference on Industrial Electronics and Applications*, Hangzhou, China, 2014

Chapter 2
Dependent Component Analysis Exploiting Nonnegativity and/or Time-Domain Sparsity

Abstract It is well-known that many real-world signals are nonnegative [1–8], i.e., their sample values are either zero or greater than zero, such as images. Obviously, nonnegativity is different from the statistical information of sources. Depending on the kinds of dependent sources, the nonnegativity of the source signals could be exploited to carry out dependent component analysis (DCA), i.e., separate these unknown dependent sources from their observed mixtures. If the sources also have certain level of sparsity in time domain, then the nonnegativity and time-domain sparsity of the source signals can be jointly employed to achieve DCA. In this chapter, three classes of dependent component analysis methods are introduced and analyzed, which are the nonnegative sparse representation (NSR) based methods, the convex geometry analysis (CGA) based methods, and the nonnegative matrix factorization (NMF) based methods. These methods either exploit the nonnegativity of the sources or both the nonnegativity and time-domain sparsity of the sources.

Keywords Nonnegative matrix factorization · Sparse representation · Convex geometry analysis

2.1 Nonnegative Sparse Representation Based Methods

Nonnegativity and sparsity constraints appear in various signal decomposition problems. For instance, in image processing, nonnegative sparse decomposition is related to the extraction of relevant parts from the images whose variables and parameters correspond to pixels [4]; in machine learning, sparseness is closely related to feature selection in learning algorithms, while nonnegativity relates to probability distributions [1]; in environmental science, scientists investigate a relative proportion of different pollutants in water or air, where proportion coefficients are nonnegative and the distributions of pollutants are often sparse [1]. Thus, it is a natural choice of applying NSR to DCA. We start from investigating the sparsity measures for nonnegative signals.

© The Author(s) 2015
Y. Xiang et al., *Blind Source Separation*,
SpringerBriefs in Signal Processing, DOI 10.1007/978-981-287-227-2_2

2.1.1 Sparsity Measures for Nonnegative Signals

A number of functions have been designed to measure the sparsity of signals. Some of them are suitable for the measurement of sparsity of a single signal and the others are for measuring the sparsity of a group of signals. For a single nonnegative signal $\mathbf{x}$, $x_i \geq 0$, $\forall i$ with n samples, one often utilizes the classic L_0-norm like [9], $L_p (0 < p < 1)$-norm like [10], and L_1-norm [11] based measures. The L_0-norm like based measure is expressed as

$$S_{\mathbf{x}} = \#\{i | x_i \neq 0\} \tag{2.1}$$

where $\#\{i\}$ denotes the number of i. Although this measure is traditional in many mathematical settings, it is not suitable for many practical scenarios. One obvious reason is that its robustness against noise is poor. Furthermore, its derivative is zero containing no information. Thus, to find the sparsest solution, one has to employ the exhaustive search approach. This is inconvenient and costly, especially when solving large scale problems. In practice, the $L_p (0 < p < 1)$-norm like or L_1-norm based measures are often used to approximate it.

The $L_p (0 < p < 1)$-norm like based measure is as follows:

$$S_{\mathbf{x}} = \left(\sum_{i=1}^{n} x_i^p \right)^{\frac{1}{p}}. \tag{2.2}$$

This measure is a good approximation of the L_0-norm like counterpart, based on which, less observations are required to separate the sources. The L_p-norm like based measure is also widely used for signal reconstruction in the area of compressed sensing which aims to recover the original high dimensional signal from its low dimensional measurements [10].

The L_1-norm based measure is defined as

$$S_{\mathbf{x}} = \sum_{i=1}^{n} x_i. \tag{2.3}$$

In some settings, the L_1 solution can be used to find the support of the L_0 solution. Besides, the L_1 solution can be found efficiently via linear programming (LP). As a result, it is widely used to replace the L_0 based complex problems. Figure 2.1 gives a simple comparison of L_0, $L_p (p = 0.5)$, L_1 and L_2 function curves.

The above measures are very intuitive and widely used in different research areas, such as blind source separation, compressed sensing, pattern recognition, machine learning and so on. However, they are not scaled, and the corresponding quantities do not contain meaningful information. Hence, it is not convenient to use them to compare the sparsity of different signals. Concerning this problem, Hoyer develops the L_1-norm and L_2-norm based measure which is scaled from zero to one [12], and

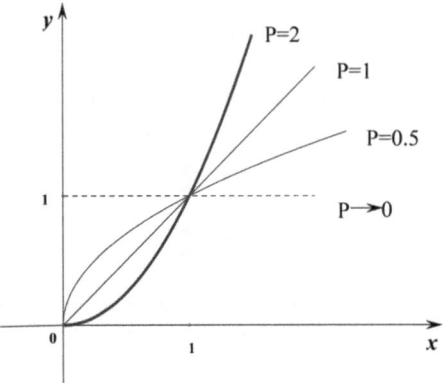

Fig. 2.1 $y = ||x||_p^p$ with different p

Yang et al. propose the following higher-order statistics based measure [4], whose value is also normalized to be in [0,1]:

$$S_{\mathbf{x}} = \frac{f_{\max} - \left(k_4 - \sigma_1 k_1^2 k_2 + \sigma_2 k_1 k_3\right)}{f_{\max} - f_{\min}} \tag{2.4}$$

where $\sigma_1 > 0$ and $\sigma_2 = (2\sigma_1 - 4)/3$ are two bounded constants, and $f_{\min} = (1 - \sigma_1 + \sigma_2)k_1^4$, $f_{\max} = \left((1/n^3) - (\sigma_1/n) + (\sigma_2/n^2)\right)k_1^4$, and $k_i = ||\mathbf{x}||_i^i$, $i = 1, 2, 3, 4$. Here, n is the number of samples. In the case of $\sigma_1 = 2$, it is easy to derive that $\sigma_2 = 0$, $f_{\min} = -k_1^4$ and $f_{\max} = (1/n^3 - 2/n)k_1^4$. Then it results from (2.4) that

$$S_{\mathbf{x}} = \frac{(\frac{1}{n^3} - \frac{2}{n})k_1^4 - k_4 + 2k_1^2 k_2}{(\frac{1}{n^3} - \frac{2}{n} + 1)k_1^4}. \tag{2.5}$$

For the purpose of visual comparison, we use the statistics based sparsity measure in [4] to compute the $S_{\mathbf{x}}$ values of three signals with different sparsity and the result is shown in Fig. 2.2. It can be seen that the sparsity measure in [4] matches well with the real sparsity of these signals.

As for the sparsity measure of a group of m-dimensional nonnegative signals $\mathbf{X} = [\mathbf{x}_1^T, \ldots, \mathbf{x}_r^T]^T$, a simple scheme is to first vectorize them and then use some existing sparsity measure. For example, if we use Hoyer's approach in [12], the sparsity measure for $\mathbf{X}$ can be described as

$$S_{\mathbf{X}} = \frac{\sqrt{mr} - \left(\sum\limits_{i=1}^{m}\sum\limits_{j=1}^{r} x_{ij}\right) \Big/ \left(\sqrt{\sum\limits_{i=1}^{m}\sum\limits_{j=1}^{r} x_{ij}^2}\right)}{\sqrt{mr} - 1}. \tag{2.6}$$

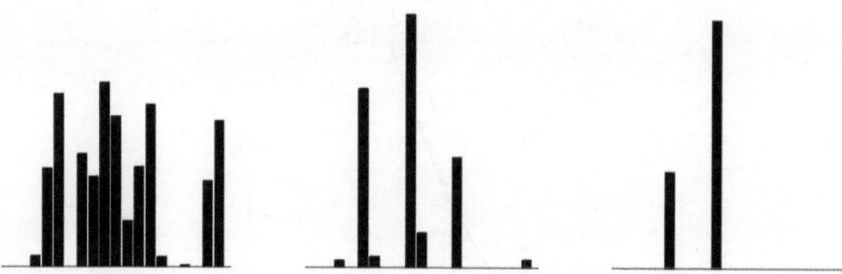

Fig. 2.2 Illustration of various degrees of sparseness. According to (2.5), the S_x values corresponding to the three signals (from left to right) are 0.1, 0.5, 0.9, respectively

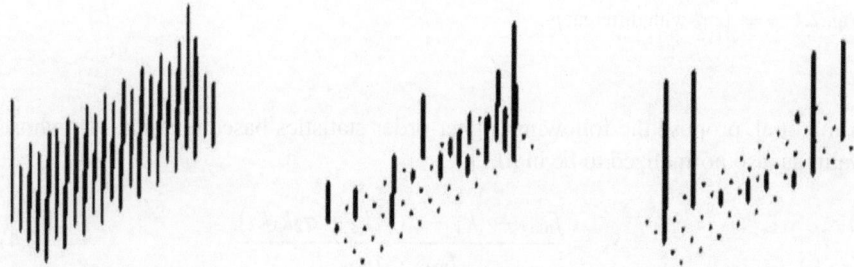

Fig. 2.3 Illustration of different degrees of sparseness. According to (2.7), the S_X values corresponding to the three 2-D signals (from left to right) are 0.1, 0.5, 0.9, respectively

The larger the index S_X, the sparser the matrix $\mathbf{X}$. In the case that each row of $\mathbf{X}$ satisfies sum-to-one, one can also use the following determinant based measure [13]:

$$S_{\mathbf{X}} = \det\left(\mathbf{XX}^T\right). \qquad (2.7)$$

Figure 2.3 gives an illustration of three 2-D signals with different levels of sparsity measured by (2.7).

There also exist some other useful sparsity measures, such as the L_0^{ε} measure [14], the $\tanh_{a,b}$ measure [15], the log measure [16], the kurtosis k_4 measure [17], the Gaussian entropy diversity measure H_G, the Shannon entropy diversity measure H_S [18], the pq-mean measure [19], and the following Gini-curve based measure which is originally used to measure the inequality of wealth in economics [16]:

$$S_{\mathbf{x}} = 1 - 2\sum_{i=1}^{n} \frac{x_i}{||\mathbf{x}||_1}\left(\frac{n - i + \frac{1}{2}}{n}\right) \qquad (2.8)$$

Table 2.1 Commonly used sparsity measures

	Measure function
L_0	$-\#\{i\|x_i \neq 0\}$
L_0^{ε}	$-\#\{i\|x_i > \varepsilon\}$
$-L_1$	$-\sum\limits_{i=1}^{n} x_i$
$-L_p$	$-(\sum\limits_{i=1}^{n} x_i^p)^{\frac{1}{p}},\ 0 < p < 1$
$-\frac{L_2}{L_1}$	$-\dfrac{\sqrt{\sum\limits_{i=1}^{n} x_i^2}}{\sum\limits_{i=1}^{n} x_i}$
$-\tanh_{a,b}$	$-\sum\limits_{i=1}^{n} \tanh((ax_i)^b)$
$-\log$	$-\sum\limits_{i=1}^{n} \log(1 + x_i^2)$
κ_4	$\dfrac{\sum\limits_{i=1}^{n} x_i^4}{(\sum\limits_{i=1}^{n} x_i^2)^2}$
H_G	$-\sum\limits_{i=1}^{n} \log x_i^2$
H_S	$-\sum\limits_{i=1}^{n} \tilde{x}_i \log \tilde{x}_i^2,\ \tilde{x}_i = \dfrac{x_i^2}{\|\mathbf{x}\|_2^2}$
Hoyer	$(\sqrt{x} - \dfrac{\sum\limits_{i=1}^{n} x_i}{\sqrt{\sum\limits_{i=1}^{n} x_i^2}})(\sqrt{n} - 1)^{-1}$
pq-mean	$-(\frac{1}{n}\sum\limits_{i=1}^{n} x_i^p)^{\frac{1}{p}}(\frac{1}{n}\sum\limits_{i=1}^{n} x_i^q)^{-\frac{1}{q}},\ p < q$
Gini	$1 - 2\sum\limits_{i=1}^{n} \frac{x_{(i)}}{\|\mathbf{x}\|_1}(\frac{n-i+\frac{1}{2}}{n}),\ x_{(1)} \leq x_{(2)} \leq \cdots \leq x_{(n)}$

where the elements of $\mathbf{x}$ are with ascending order, i.e., $x_1 \leq x_2 \leq \cdots \leq x_n$. Table 2.1 shows a list of commonly used sparsity measures, where the functions are modified such that larger measures correspond to sparser signals and some of them are also shown in [20].

2.1.2 Estimation of Mixing Matrix and Source Signals

Consider the mixing system model $\mathbf{Y} = \mathbf{AX}$ shown in (1.6) and assume that the source signals $\mathbf{X}$ are nonnegative and sparse. Based on this mixing system model, some NSR methods have been proposed for different scenarios, including the quadratic programming (QP) based method for the determined mixing system [13] and the clustering based method for the underdetermined mixing case [21]. We will discuss the determined and underdetermined scenarios separately.

2.1.2.1 Determined Mixing System

In the determined mixing system, the number of the observations are equal or greater than that of the sources. Similar to the independence based method for BSS, one can implement DCA by finding a separation matrix $\mathbf{B}$ such that the product $\mathbf{BA}$ is a permutation matrix neglecting the inherent scaling issue. Since the sources are nonnegative and sparse, one can utilize the following optimization model, which exploits the source sparsity based on the determinant measure, to obtain the separation matrix [13]:

$$\text{Maximize}: \quad \det(\mathbf{BYY}^T\mathbf{B}^T) \tag{2.9}$$

$$\text{s.t.} \begin{cases} \sum_{j=1}^{m} b_{ij}y_{jt} \geq 0, \forall i, t \\ \sum_{j=1}^{m} b_{ij} = 1, \forall i \end{cases}$$

where $\mathbf{Y}$ is normalized to be row-sum-to-one in prior. The objective function reflects the sparsity of the estimated sources. Regarding the constraints, the first one denotes the nonnegativity and the second one is used to scale each estimated source to be sum-to-one.

In order to solve the model (2.9), the iterative sparseness maximization based on quadratic programming (ISM-QP) algorithm is developed in [13], where the separation matrix is optimized row-by-row iteratively. For the ith row $\bar{\mathbf{b}}_i$ of $\mathbf{B}$, the following optimization model is further derived:

$$\text{Maximize}: \quad \bar{\mathbf{b}}_i\mathbf{C}\bar{\mathbf{b}}_i^T \tag{2.10}$$

$$\text{s.t.} \begin{cases} \sum_{j=1}^{m} b_{ij}y_{jt} \geq 0, \ \forall i, t \\ \sum_{j=1}^{m} b_{ij} = 1 \end{cases}$$

where $\mathbf{C}$ is a matrix independent of $\bar{\mathbf{b}}_i$. Specifically,

$$\mathbf{C} = \mathbf{C}_1 + \mathbf{C}_2 + \mathbf{C}_3 \tag{2.11}$$

with

$$
\begin{cases}
\mathbf{C}_1 = \tilde{\mathbf{X}} \sum_{j=1}^{i-1} (-1)^{i+j} \overline{\mathbf{b}}_j^T \left[\sum_{t=1}^{i-1} (-1)^{t+i-1} \det\left(\tilde{\mathbf{Y}}_{ij,t(i-1)}\right) \overline{\mathbf{b}}_t \right. \\
\qquad \left. + \sum_{t=i}^{n-1} (-1)^{t+i-1} \det\left(\tilde{\mathbf{Y}}_{ij,t(i-1)}\right) \overline{\mathbf{b}}_{t+1} \right] \tilde{\mathbf{X}} \\
\mathbf{C}_2 = (-1)^{i+i} \det\left(\tilde{\mathbf{Y}}_{ii}\right) \tilde{\mathbf{X}} \\
\mathbf{C}_3 = \tilde{\mathbf{X}} \sum_{j=i+1}^{n} (-1)^{i+j} \overline{\mathbf{b}}_j^T \left[\sum_{t=1}^{i-1} (-1)^{t+i} \det\left(\tilde{\mathbf{Y}}_{ij,ti}\right) \overline{\mathbf{b}}_t \right. \\
\qquad \left. + \sum_{t=i}^{n-1} (-1)^{t+i} \det\left(\tilde{\mathbf{Y}}_{ij,ti}\right) \overline{\mathbf{b}}_{t+1} \right] \tilde{\mathbf{X}}
\end{cases}
$$

where $\tilde{\mathbf{X}} = \mathbf{Y}\mathbf{Y}^T$, $\tilde{\mathbf{Y}} = \mathbf{B}\mathbf{Y}\mathbf{Y}^T\mathbf{B}^T$, and $\tilde{\mathbf{Y}}_{ij}$ denotes a $(r-1) \times (r-1)$ submatrix of $\tilde{\mathbf{Y}}$ with the ith row and the jth column removed. By analyzing the inequalities in the constraints, the model (2.10) can be rewritten as

$$
\text{Maximize}: \quad \overline{\mathbf{b}}_i \mathbf{C} \overline{\mathbf{b}}_i^T \tag{2.12}
$$

$$
\text{s.t.} \begin{cases} \sum_{j=1}^{m} b_{ij} v_{jl} \geq 0, \ \forall l \in \{1, 2, \ldots, L\} \\ \sum_{j=1}^{m} b_{ij} = 1 \end{cases}
$$

where $\forall i$, $\mathbf{v}_i = \sum_{j=1}^{m} v_{ji}$ denotes the ith extreme point of the convex hull spanned by the observations.

The ISM-QP algorithm for solving DCA with r sources from the observation $\mathbf{Y}$ is summarized in Table 2.2 below. Some highly correlated face images are used to test the effectiveness of this algorithm and Fig. 2.4 shows the separation results. We can see that the ISM-QP algorithm achieves almost perfect recoveries.

Table 2.2 ISM-QP algorithm [13]

Step 1 (Preprocessing)	Normalize each row of $\mathbf{Y}$ to be sum-to-one and find the extreme points $\mathbf{v}_1, \mathbf{v}_2, \ldots, \mathbf{v}_L$ of the convex hull spanned by $\mathbf{Y}$
Step 2 (Initialization)	Let $i = 1$ and set an initial matrix with row-sum-to-one for $\mathbf{B}$
Step 3 (Iteration)	(i) Obtain the optimal solution $\overline{\mathbf{b}}_i^\star$ of (2.12)
	(ii) Set $i = i + 1$ until a given stop criterion is satisfied
	(iii) If $i > r$, reset $i = \mathrm{mod}(i, r)$, and if $i = 0$, set $i = r$
Step 4 (Estimation)	The source matrix is estimated by $\mathbf{B}^\star\mathbf{Y}$, where $\mathbf{B}^\star = [(\overline{\mathbf{b}}_1^\star)^T, \ldots, (\overline{\mathbf{b}}_r^\star)^T]^T$

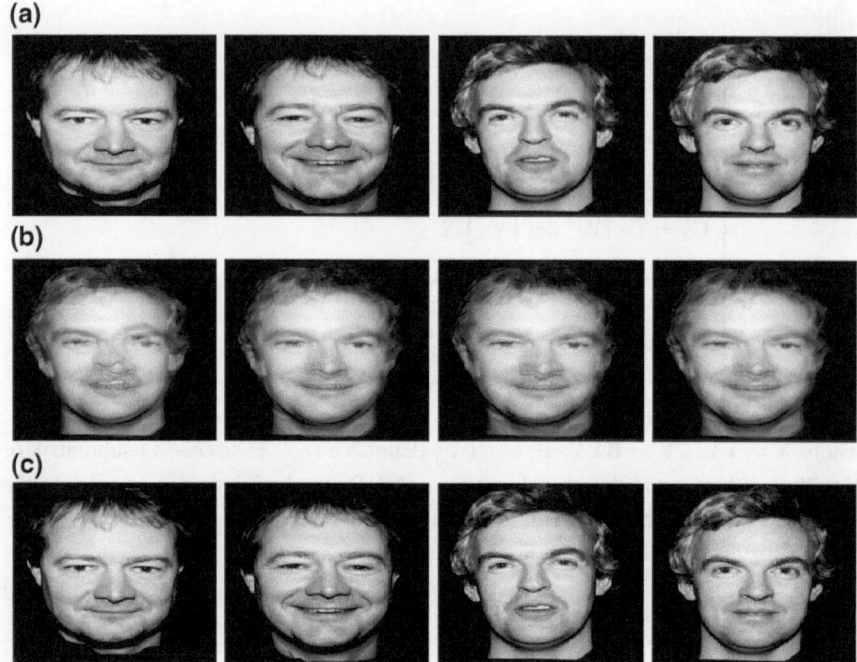

Fig. 2.4 Results of the ISM-QP algorithm in separating correlated face images. **a** Four correlated source images; **b** Four mixtures using random mixing matrix; **c** Four recoveries using ISM-QP

2.1.2.2 Underdetermined Mixing System

In the underdetermined mixing system scenario, the number of sources is greater than that of the observations. Since the mixing matrix is not invertible, it is impossible to implement DCA by searching a separation matrix. A feasible way of solving this challenging problem is to employ a two-step scheme: first estimate the mixing matrix **A** and then recover the sources **X**. Regarding the estimation of the mixing matrix, a popular approach is the clustering based one, which assumes that the sources are sparse [21]. After the mixing matrix is obtained, the estimation of the sources falls into the sparse reconstruction problem. In order to directly utilize the existing sparse representation algorithms, one can recover the source matrix column by column. In this case, the widely used L_1-norm based method is a good option [11]. Furthermore, under some conditions, the subspace based scheme gives a more efficient way to conduct source recovery [21]. Table 2.3 shows the detailed structure of the subspace based algorithm.

Table 2.3 Estimating the source matrix **X** [21]

Step 1 (Identification)	Calculate the set of k-codimensional subspaces $\mathbb{H}$ produced by taking the linear hull of every subset of the columns of **A** with $m - 1$ elements
Step 2 (Iteration)	For $i = 1, \ldots, n$ (i) identify the space $H \in \mathbb{H}$ containing $\mathbf{y}_i$, and project $\mathbf{y}_i$ onto H to $\tilde{\mathbf{y}}_i$; (ii) if H is produced by the linear hull of column vectors $\mathbf{a}_{i_1}, \ldots,$ $\mathbf{a}_{i_{m-1}}$, then find coefficients λ_{ij} such that $$\tilde{\mathbf{y}}_i = \sum_{j=1}^{m-1} \lambda_{ij} \mathbf{a}_{i_j}$$
Step 3 (Estimation)	The estimation of $\mathbf{x}_i, \forall i$ contains λ_{ij} in the place i_j for $j = 1, \ldots, m-1$, and its remaining components are zero

2.1.3 Uniqueness Conditions

Regarding the mixing model $\mathbf{Y} = \mathbf{AX}$ with $\mathbf{A} \in \mathbb{R}^{m \times r}$ and $\mathbf{X} \in \mathbb{R}^{r \times n}$, the uniqueness conditions of NSR are related to both the mixing matrix **A** and the source matrix **X**. In the case that **A** is determined, i.e., $m \geq r$, we have the following theorem:

Theorem 2.1 *If the mixing matrix* **A** *is full column rank, the source matrix* **X** *is nonnegative, and there exists an $r \times r$ submatrix* $\hat{\mathbf{X}}$ *satisfying* $\det(\hat{\mathbf{X}}\hat{\mathbf{X}}^T) = 1$, *where* $\hat{\mathbf{X}}$ *is normalized to be row-sum-to-one, then it holds that* [13]

$$\mathbf{B}^\star \mathbf{A} = \mathbf{P} \tag{2.13}$$

where $\mathbf{B}^\star$ *is the optimal solution of* (2.9) *and* **P** *is a permutation matrix.*

When the mixing system is underdetermined, i.e., $m < r$, the uniqueness analysis becomes much more complex. If **A** is unknown, the following theorem gives the sufficient conditions:

Theorem 2.2 *Assume that $m \leq r \leq n$, any $m \times m$ square submatrix of* $\mathbf{A} \in \mathbb{R}^{m \times r}$ *is nonsingular,* $\mathbf{X} \in \mathbb{R}^{r \times n}$ *is sufficiently rich and its each column has at most $m - 1$ nonzero elements, then the matrix* $\mathbf{Y} \in \mathbb{R}^{m \times n}$ *can be represented uniquely in the form* $\mathbf{Y} = \mathbf{AX}$ *if the following conditions are satisfied* [21]:

(i) *the columns of* **Y** *lie in the union $\mathbb{H}$ of C_r^{m-1} different hyperplanes, each column lies in only one such hyperplane, each hyperplane contains at least m columns of* **Y** *such that each $m - 1$ of them are linearly independent;*

(ii) *for each $i \in 1, \ldots, r$, there exist $p = C_{r-1}^{m-2}$ different hyperplanes $\{H_{i,j}\}_{j=1}^p$ in $\mathbb{H}$ such that their intersection $L_i = \cap_{j=1}^p \{H_{i,j}\}$ is 1-D subspace;*

(iii) *any m different L_i span the whole $\mathbb{R}^m$.*

Sometimes, the mixing matrix is known or can be calculated by some methods. In this case, the uniqueness of NSR is related to the uniqueness of nonnegative solution in a underdetermined system. We have the following theorem:

Theorem 2.3 *Given that* y *and* $\mathbf{A} \in \mathbb{R}^{m \times r}$ *(where* $m < r$*) for the system* $\mathbf{y} = \mathbf{A}\mathbf{x}$ *(where* $\mathbf{x} \succeq \mathbf{0}$*) with finite solutions. Let* $\hat{\mathbf{x}}$ *be a solution to this problem, it is the unique solution if* $\hat{\mathbf{x}}$ *satisfies* [22]

$$||\hat{\mathbf{x}}||_0 < \frac{1}{2t_{\mathbf{A}}} \tag{2.14}$$

where $||\hat{\mathbf{x}}||_0$ *denotes the number of the non-zero element of* $\hat{\mathbf{x}}$, $t_{\mathbf{A}} = \rho(\mathbf{A})/(1+\rho(\mathbf{A}))$ *and* $\rho(\mathbf{A})$ *denotes the one-sided coherence which is defined as*

$$\rho(\mathbf{A}) = \max_{i,j;j \neq i} \frac{|\mathbf{a}_i^T \mathbf{a}_j|}{||\mathbf{a}_i||_2^2}. \tag{2.15}$$

From the viewpoint of blind source separation, the above theorem shows the condition of uniquely recovering one column of $\mathbf{X}$. Also, it can be easily extended to the following corollary:

Corollary 2.1 *Given that* $\mathbf{Y} \in \mathbb{R}^{m \times n}$ *and* $\mathbf{A} \in \mathbb{R}^{m \times r}$ *with* $m < r$. *If* $\forall i \in \{1, \ldots, n\}$, *the solution* $\hat{\mathbf{x}}_i$ *of* $\mathbf{y}_i = \mathbf{A}\mathbf{x}_i$ *satisfies*

$$||\hat{\mathbf{x}}_i||_0 < \frac{1}{2t_{\mathbf{A}}} \tag{2.16}$$

then $\hat{\mathbf{X}} = [\hat{\mathbf{x}}_1, \ldots, \hat{\mathbf{x}}_n]$ *is the unique solution of* $\mathbf{Y} = \mathbf{A}\mathbf{X}$.

There are more uniqueness results related to the NSR problem. They range from the analysis of the nonnegative solutions to the underdetermined linear equations, including the restricted isometry property related conditions [23], the k-neighborly features [24], etc.

2.2 Convex Geometry Analysis Based Methods

In the context of nonnegative sources, there might be some geometric structures in the observations and the sources. For example, the biomedical image and human portraits are often with *local dominance* feature, under which the source signals correspond to the extreme points of some observation-constructed convex polyhedral set [5]; the hyper-spectral image abundance (or source) matrix has column-sum-to-one feature, such that it corresponds to the minimum volume simplex among those enclose of the observed data [6]; and the dynamic positron emission tomography images and the mass spectra for metabolomics are often the minimum aperture simplicial convex cones which contain their respective mixtures [7]. The use of these geometric features, instead of the statistical features of the sources, can facilitate the blind separation of mutually correlated sources.

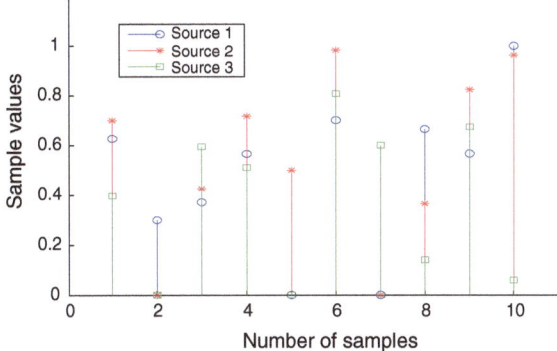

Fig. 2.5 Illustration of three sources with local dominance feature, where the dominant indices are 2, 5, 7 respectively

2.2.1 Geometric Features

Local dominance, also called pure source sample in [25], is an important geometric feature existing in some sources. It means that for each source there is at least one time instant at which the source dominates. A mathematical definition of local dominance is as follows:

Definition 2.1 (*Local dominance*): A group of sources $\bar{\mathbf{x}}_1^T, \ldots, \bar{\mathbf{x}}_r^T$ have local dominance feature if for each $i \in \{1, \ldots, r\}$, there exists an index l_i such that $\bar{\mathbf{x}}_i(l_i) > 0$ and $\bar{\mathbf{x}}_j(l_i) = 0, \forall j \neq i$.

Figure 2.5 gives an illustration of three sources with local dominance feature, where the dominant indices for the three sources are 2, 5, 7 respectively.

The local dominance feature may be completely satisfied or serve as a good approximation when the source signals are sparse (or contain many zeros). For example, in brain magnetic resonance imaging (MRI), the nonoverlapping region of the spatial distribution of fast perfusion and slow perfusion source images can be larger than 95 % [27]. In the hyperspectral unmixing problem, the abundances of the ground covers are often quite sparse, and thus the source images (corresponding to the abundances) tend to satisfy local dominance [26]. It may also be appropriate to consider this feature when the source signals exhibit high contrast, which could exist in sources such as face images and natural images. The local dominance feature is widely applied to solving the BSS problem [5, 28].

Another geometric feature is the minimum cone feature related to the mixing matrix [25, 29, 30]. From the mixing model $\mathbf{Y} = \mathbf{AX}$ with $\mathbf{X} \succeq \mathbf{0}$, we can see that each column of $\mathbf{Y}$ is the nonnegative linear combination of the columns of $\mathbf{A}$. This implies that the cones which enclose $\mathbf{Y}$ are related to $\mathbf{A}$. Specifically, it is found that the vertices of the simplicial cone and convex hull (defined below) with minimum volume correspond to the columns of $\mathbf{A}$ under some conditions.

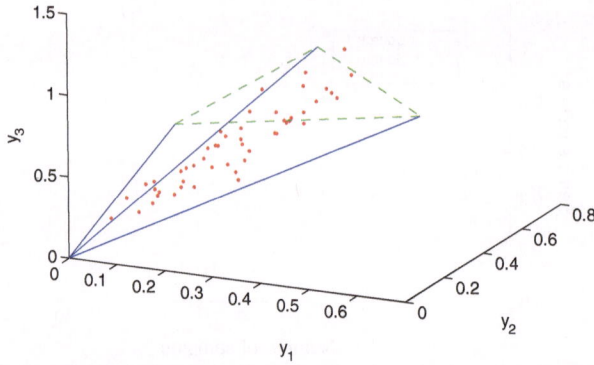

Fig. 2.6 Scatter plot of mixed data included in **Span**$^+$(**A**)

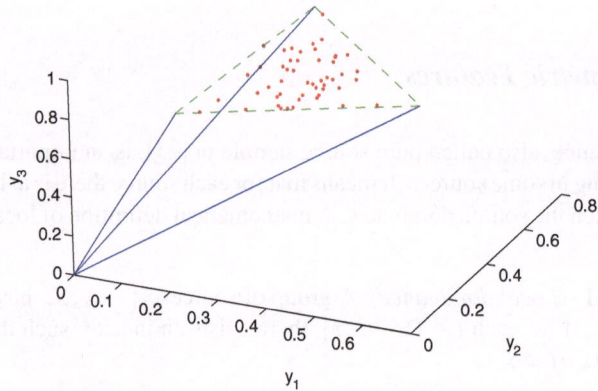

Fig. 2.7 Scatter plot of mixed data included in **Conv**$^+$(**A**)

Definition 2.2 (*Simplicial cone*): The simplicial cone generated by the columns of **A**, denoted by **Span**$^+$(**A**), is defined as [7]:

$$\mathbf{Span}^+(\mathbf{A}) = \{\mathbf{y}|\mathbf{y} = \mathbf{Ax} \text{ with } \mathbf{x} \in \mathbb{R}_+^r\}. \tag{2.17}$$

Definition 2.3 (*Convex hull*): The convex hull generated by the columns of **A**, denoted by **Conv**$^+$(**A**), is defined as [6]:

$$\mathbf{Conv}^+(\mathbf{A}) = \{\mathbf{y}|\mathbf{y} = \mathbf{Ax}, \ \sum_{i=1}^{r}\mathbf{x}_i = 1, \ \mathbf{x}_i \in \mathbb{R}_+^r\}. \tag{2.18}$$

Figures 2.6 and 2.7 illustrate respectively **Span**$^+$(**A**) and **Conv**$^+$(**A**) in the case $r = 3$, where **A** is randomly generated as

$$\mathbf{A} = \begin{bmatrix} 0.6493 & 0.1765 & 0.2609 \\ 0.2088 & 0.1159 & 0.6469 \\ 0.9641 & 0.8015 & 0.9105 \end{bmatrix}.$$

2.2.2 Estimation of Source Signals

There are many CGA based methods for DCA, which explicitly exploit the local dominance feature of the sources, including the convex analysis of mixtures of non-negative sources using linear programming (CAMNS-LP) [5], the project pursuit (PP) [28] , the vertex component analysis (VCA) [26] and the modified VCA [31]. Here, we introduce the first two algorithms for reference. Regarding the CAMNS-LP method, it combines the convex analysis and optimization techniques. In this method, one first finds, through the convex analysis, the true source signals which serve as the extreme points of some observation-constructed polyhedral set. Then, an extreme-point finding algorithm is developed, by taking advantage of the powerful tool of linear programming, for source recovery. Actually, for the given mixture $\mathbf{Y}$, it first calculates the 2-tuple $(\mathbf{C}, \mathbf{d})$ as follows:

$$\begin{cases} \mathbf{d} = \frac{1}{m} \sum_{i=1}^{m} \bar{\mathbf{y}}_i^T \\ \mathbf{C} = \left[\mathbf{q}_1(\mathbf{U}\mathbf{U}^T), \mathbf{q}_2(\mathbf{U}\mathbf{U}^T), \dots, \mathbf{q}_{r-1}(\mathbf{U}\mathbf{U}^T) \right] \end{cases} \tag{2.19}$$

where $\mathbf{U} = [\bar{\mathbf{y}}_1^T - \mathbf{d}, \dots, \bar{\mathbf{y}}_m^T - \mathbf{d}] \in \mathbb{R}^{n \times m}$, the notation $\mathbf{q}_i(\mathbf{A})$ denotes the eigenvector associated with the ith principal eigenvalue of the mixing matrix $\mathbf{A}$, and $\bar{\mathbf{y}}_i$ is the ith row of the mixture matrix $\mathbf{Y}$. This 2-tuple constructs a meaningful affine hull. Built upon this 2-tuple, a series of LP problems are constructed for separating the sources. Then, the CAMNS-LP method recovers the sources iteratively, where only solvable LP problems need to be processed in each iteration. Table 2.4 shows the structure of the CAMNS-LP algorithm.

As for the PP method, it first maps the observation matrix into a superplane such that one of the rows of the mapped observation matrix has equal elements with value 1. This ensures that the unaccessible source matrix is normalized to be column-sum-to-one. Then, based on the property of the normalized source matrix, it estimates one column of the mixing matrix $\mathbf{A}$ by searching an optimal projection vector for the mapped observation matrix. After that, it estimates another column by searching another optimal vector in the subspace orthogonal to the already estimated columns. All columns of the mixing matrix $\mathbf{A}$ can be obtained by repeating this process. The PP method works under the same conditions as those of the CAMNS-LP method but it has much less computational complexity as it only needs to solve one LP problem [28]. The PP algorithm is summarized in Table 2.5.

Table 2.4 CAMNS-LP algorithm [5]

Step 1 (Preprocessing)	Calculate the 2-tuple $(\mathbf{C}, \mathbf{d})$ of the given $\mathbf{Y}$ by (2.19)
Step 2 (Initialization)	Set $l = 0$ and $\mathbf{B} = \mathbf{I}_n$, where $\mathbf{I}_n$ is the $n \times n$ identity matrix
Step 3 (Iteration)	While $l \leq r$, (i) let $\mathbf{h} = \mathbf{Bw}$, where $\mathbf{w} \sim N(0, 1)$ is a randomly generated vector; (ii) solve the LPs $$p^\star = \min_{\alpha:\mathbf{C}\alpha+\mathbf{d}\geq 0} \mathbf{h}^T(\mathbf{C}\alpha + \mathbf{d})$$ $$q^\star = \max_{\alpha:\mathbf{C}\alpha+\mathbf{d}\geq 0} \mathbf{h}^T(\mathbf{C}\alpha + \mathbf{d})$$ and obtain their optimal solutions, denoted by $\alpha_1^\star$ and $\alpha_2^\star$, respectively; (iii) if $l = 0$, let $$\widehat{\mathbf{S}} = [\mathbf{C}\alpha_1^\star + \mathbf{d}, \mathbf{C}\alpha_2^\star + \mathbf{d}]$$ else, update $\widehat{\mathbf{S}}$ by $$\widehat{\mathbf{S}} := \begin{cases} [\widehat{\mathbf{S}}, \mathbf{C}\alpha_1^\star + \mathbf{d}], & \text{if } |p^\star| \neq 0 \\ [\widehat{\mathbf{S}}, \mathbf{C}\alpha_2^\star + \mathbf{d}], & \text{if } |q^\star| \neq 0 \end{cases};$$ (iv) update l to be the number of columns of $\widehat{\mathbf{S}}$, and apply QR decomposition to $\widehat{\mathbf{S}}$ as $$\widehat{\mathbf{S}} = \mathbf{Q}_l\mathbf{R}_l$$ where $\mathbf{Q}_l \in \mathbb{R}^{n \times l}$ and $\mathbf{R}_l \in \mathbb{R}^{l \times l}$; (v) update $\mathbf{B}$ by $$\mathbf{B} := \mathbf{I}_n - \mathbf{Q}_l\mathbf{Q}_l^T$$
Step 4 (Estimating $\mathbf{X}$)	Finally, the source matrix $\mathbf{X}$ is estimated by $\widehat{\mathbf{X}} = \widehat{\mathbf{S}}^T$

To give a visual comparison of the performance of the PP and CAMNS-LP algorithms, we use them to test four correlated fingerprint images,[1] where the correlation coefficient matrix $\mathbf{C}$ is:

$$\mathbf{C} = \begin{bmatrix} 1.0000 & 0.7908 & 0.5939 & 0.6965 \\ 0.7908 & 1.0000 & 0.6548 & 0.7712 \\ 0.5939 & 0.6548 & 1.0000 & 0.7767 \\ 0.6965 & 0.7712 & 0.7767 & 1.0000 \end{bmatrix}.$$

The mixing matrix $\mathbf{A}$ is generated randomly as:

$$\mathbf{A} = \begin{bmatrix} 0.7814 & 0.4464 & 0.3072 & 0.3298 \\ 0.4157 & 0.5367 & 0.2705 & 0.3822 \\ 0.4703 & 0.7291 & 0.6629 & 0.4115 \\ 0.4970 & 0.3533 & 0.5180 & 0.9035 \end{bmatrix}.$$

Figure 2.8 shows the source images, the mixtures, and the recoveries by using the PP algorithm and the CAMNS-LP algorithm, respectively. It can be seen that both of

---

[1] See http://biometrics.cse.msu.edu/fvc04db/index.html.

Table 2.5 PP algorithm [28]

Step 1 (Preprocessing)	(i) Obtain $\mathbf{u}$ satisfying $\mathbf{u}^T \mathbf{y}_i > 0$, $\forall i$ and suppose $u_q \neq 0$
	(ii) Compute $\mathbf{D}$ by $$\mathbf{D} = \text{diag}(\mathbf{1}^T \oslash (\mathbf{u}^T \mathbf{Y}))$$ where $\mathbf{1}$ is a all-1 column vector and $\oslash$ denotes component-wise division
	(iii) Let $\tilde{\mathbf{I}}_m$ be the $m \times m$ identity matrix with the qth row replaced by $\mathbf{u}^T$. Map $\mathbf{Y}$ into $\tilde{\mathbf{Y}}$ by $$\tilde{\mathbf{Y}} = \tilde{\mathbf{I}}_m \mathbf{Y} \mathbf{D}$$
Step 2 (Estimating $\mathbf{a}_1$)	(i) Set $\mathbf{v} = \mathbf{0}$ and generate randomly a full-rank square matrix $\mathbf{B}$
	(ii) Update $\mathbf{v}$ using the scheme (related to $\mathbf{B}$) in [28] and estimate $\hat{\mathbf{a}}_1$ by $$\hat{\mathbf{a}}_1 = \mathbf{y}_j$$ where $$j = \begin{cases} \arg\max(\mathbf{v}^T \tilde{\mathbf{Y}}), & \text{if } \max(\mathbf{b}_1^T \tilde{\mathbf{Y}}) > 0 \\ \arg\min(\mathbf{v}^T \tilde{\mathbf{Y}}), & \text{else} \end{cases}$$
Step 3 (Estimating $\mathbf{a}_2, \ldots, \mathbf{a}_r$)	For $k = 1, 2, \ldots, r-1$,
	(i) update $\hat{\mathbf{A}}_k$ and $\hat{\mathbf{A}}_k^{\perp}$ by $$\begin{cases} \hat{\mathbf{A}}_k = [\hat{\mathbf{a}}_1, \ldots, \hat{\mathbf{a}}_k] \\ \hat{\mathbf{A}}_k^{\perp} = \left(\mathbf{I}_r - \hat{\mathbf{A}}_k (\hat{\mathbf{A}}_k^T \hat{\mathbf{A}}_k)^{-1} \hat{\mathbf{A}}_k^T \right) \mathbf{H} \end{cases}$$ where $\mathbf{H} \in \mathbb{R}^{r \times (r-k)}$ is a matrix of full column rank;
	(ii) update $\mathbf{B}$ by $$\begin{cases} \mathbf{B}(1:r, 1:r-k) = \hat{\mathbf{A}}_k^{\perp} \\ \mathbf{B}(1:r, r-k+1:r) = \hat{\mathbf{A}}_k \end{cases}$$
	(iii) estimate $\hat{\mathbf{a}}_{k+1}$ using the method shown in Step 2.
Step 4 (Estimating $\mathbf{X}$)	Let $\hat{\mathbf{A}}_r = [\hat{\mathbf{a}}_1, \hat{\mathbf{a}}_2, \ldots, \hat{\mathbf{a}}_r]$, then the source matrix is estimated by $$\hat{\mathbf{X}} = \hat{\mathbf{A}}_r^{-1} \mathbf{Y}$$

them achieve satisfactory separating results. The corresponding CPU running times for these two algorithms are 2.9172 and 40.0455 s, respectively, indicating that the PP algorithm is much faster than the CAMNS-LP method.

Also, there are some methods which do not require the local dominance condition, such as the minimum volume simplex (MVS) [6] and the simplicial cone shrinking algorithm (SCSA) [7]. The MVS algorithm assumes that the sources are column-sum-to-one, i.e., the full additivity. In the noiseless case, one can relax the full additivity assumption by normalizing each column of the data matrix to a unit sum. However, in the noisy case, enforcing this normalization may amplify noise and thus yield a bad estimation of the sources, especially if the number of sources is overestimated. Different from the MVS algorithm, SCSA estimates the mixing matrix and the sources by finding the minimum aperture simplicial cone (MASC) containing the scatter plot of the mixed data. It needs neither the local dominance condition nor the full additivity assumption, applicable to a wider range of applications. Generally,

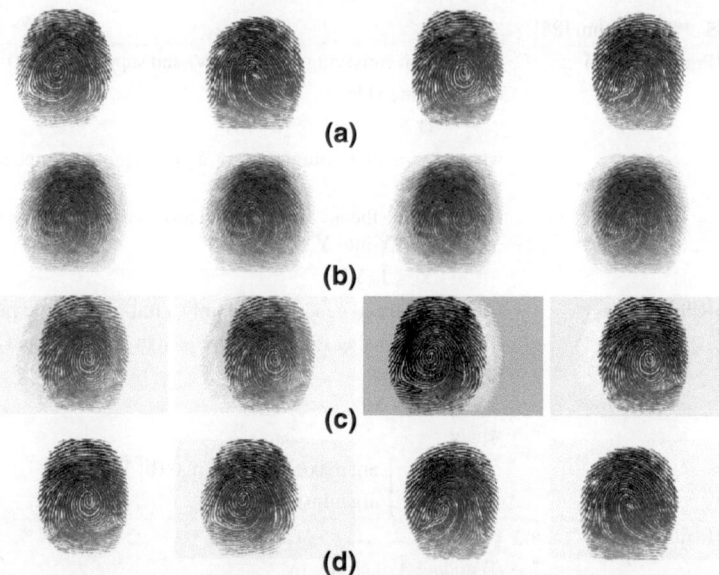

Fig. 2.8 Results of separating correlated fingerprint images by the PP and CAMNS-LP algorithms.
a Four correlated source images; **b** Four mixtures; **c** Four recoveries using PP; **d** Four recoveries
using CAMNS-LP

SCSA first finds a proper initial simplicial cone by using the VCA algorithm in [26],
then decreases the aperture of the current simplicial cone iteratively. A summary of
SCSA is shown in Table 2.6.

2.2.3 Source Identifiability Analysis

Regarding the source identifiability issue in relation to the mixing model $\mathbf{Y} = \mathbf{AX}$,
there are several conclusions shown in the following theorems.

Theorem 2.4 *Assuming that the sources are nonnegative with local dominance feature and the mixing matrix is full column rank with row-sum-to-one, the extreme
points of the following polyhedral set correspond to the r true source vectors [5]:*

$$\{\mathbf{y} \in \mathbb{R}^n | \mathbf{y} = \mathbf{C}\alpha + \mathbf{d} \succeq \mathbf{0}, \alpha \in \mathbb{R}^{r-1}\} \tag{2.20}$$

where $(\mathbf{C}, \mathbf{d})$ *is obtained from* $\mathbf{Y}$ *by* (2.19).

This theorem shows how the local dominance feature affects the identification of the
sources. If the mixing system satisfies the mentioned conditions, the sources can be
recovered by searching the extreme points related to (2.20).

Table 2.6 A summary of SCSA [7]

Step 1 (Initialization)	Set $\mathbf{W} = \mathbf{I}_m$ or find it by using the VCA method and set $\mathbf{D} = \mathbf{W}^{-1}\mathbf{Y}$
Step 2 (Iteration)	(i) $\forall i \in \{1, \ldots, r\}$, compute $\mathbf{V}_i$ by $$\mathbf{V}_i = \begin{bmatrix} 1 & 0 & \ldots & 0 & v_{1i} & 0 & \cdots & 0 \\ 0 & 1 & \cdots & 0 & v_{2i} & 0 & \cdots & 0 \\ \vdots & \vdots & \ddots & \vdots & \vdots & \vdots & \ddots & \vdots \\ 0 & 0 & \cdots & 1 & v_{(i-1)i} & 0 & \cdots & 0 \\ 0 & 0 & \cdots & 0 & 1 & 0 & \cdots & 0 \\ 0 & 0 & \cdots & 0 & v_{(i+1)i} & 1 & \cdots & 0 \\ \vdots & \vdots & \ddots & \vdots & \vdots & \vdots & \ddots & \vdots \\ 0 & 0 & \cdots & 0 & v_{mi} & 0 & \cdots & 1 \end{bmatrix}$$ and update $\mathbf{W}, \mathbf{D}$ by $$\begin{cases} \mathbf{W} = \mathbf{W}\mathbf{V}_1\mathbf{V}_2 \cdots \mathbf{V}_r \\ \mathbf{D} = [\mathbf{R}_r]^{-1}[\mathbf{R}_{r-1}]^{-1} \cdots [\mathbf{R}_1]^{-1}\mathbf{D} \end{cases}.$$ (ii) Compute $\mathbf{Q}$ by $$\mathbf{Q}_{(p+1)} = \mathbf{Q}_p - \mu\left[-\frac{(\mathbf{W}^{-1})^T \mathbf{T}^{null} \mathbf{T}^T}{\sigma} + 4\gamma \ \mathbf{Q}_p(\mathbf{Q}_p^T\mathbf{Q}_p - \mathbf{I}_r)\right]$$ where μ is a learning rate parameter, $\mathbf{T} = \mathbf{W}^{-1}\mathbf{Q}_p\mathbf{Y}$, $\mathbf{T}^{null} = \exp(-\mathbf{T}/\sigma)$, $\sigma > 0$, and $\gamma \geq 0$. And let $$\begin{cases} \mathbf{W} = \mathbf{Q}\mathbf{W} \\ \mathbf{D} = \mathbf{W}^{-1}\mathbf{Q}^{-1}\mathbf{Y} \end{cases}$$ (iii) If $\mathbf{Q} = \mathbf{I}_r$, stop the iteration
Step 3 (Estimation)	$\mathbf{A}$ and $\mathbf{X}$ are estimated by using MATLAB functions $\max(\mathbf{W}, 0)$ and $\max(\mathbf{D}, 0)$, respectively

More recent results about source identification are given in [7] as follows:

Theorem 2.5 $\mathrm{Span}^+(\mathbf{A})$ *is the unique nonnegative MASC containing the scatter plot of the mixed data if and only if* $\mathrm{Span}^+(\mathbf{I}_r)$ *is the unique nonnegative MASC containing the scatter plot of the sources, where* $\mathbf{I}_r$ *denotes the* $r \times r$ *identity matrix.*

Theorem 2.6 (Necessary condition) *If* $\mathrm{Span}^+(\mathbf{I}_r)$ *is the unique nonnegative MASC containing the scatter plot of the sources, then there is at least one point of the cloud of sources on each facet of* $\mathrm{Span}^+(\mathbf{I}_r)$, *i.e.,* $\forall 1 \leq i \leq r$, $\exists k_i$ *such that* $x_i(k_i) = 0$.

Theorem 2.7 (Sufficient condition 1) *If the sources are nonnegative and locally dominant, then* $\mathrm{Span}^+(\mathbf{I}_r)$ *is the unique nonnegative MASC containing the scatter plot of sources.*

Theorem 2.8 (Sufficient condition 2) *For each facet of* $\mathrm{Span}^+(\mathbf{I}_r)$, *if at least* $r - 1$ *points of the scatter plot of the sources belong to underlined facet, and the vectors corresponding to these points are linearly independent, then* $\mathrm{Span}^+(\mathbf{I}_r)$ *is the unique nonnegative MASC containing the scatter plot of the sources.*

2.3 Nonnegative Matrix Factorization Based Methods

Like the sources, sometimes the mixing matrix is also nonnegative. For example, in remote sensing image processing, both the endmember signature matrix and the abundance matrix are nonnegative [4]. In fluorescence spectroscopy analysis, the pure species spectra and their concentrations are also nonnegative [3]. More practical mixing systems with nonnegative sources and nonnegative mixing matrix can be found in [1]. Since NMF aims to decompose a given nonnegative matrix into the product of two nonnegative matrices [8], it matches well with the BSS problem, or DCA when the sources are spatially correlated. NMF is a well developed scheme which is widely used in the areas of signal processing and pattern recognition. Over the last few years, a number of NMF based methods have been proposed to implement DCA, such as NMF-MVC [32], NMF-L1 [33], and NMF-SMC [4]. Prior to discussing these methods, we first introduce some NMF models.

2.3.1 Nonnegative Matrix Factorization Models

Assume that $\mathbf{Y}$ is a given nonnegative matrix, NMF aims to decompose $\mathbf{Y}$ into the product of two nonnegative matrices, denoted by $\mathbf{A}$ and $\mathbf{X}$, respectively. Mathematically, the standard NMF can be described as [1, 8]

$$\mathbf{Y} \approx \mathbf{AX} \tag{2.21}$$

where $\mathbf{Y} \in \mathbb{R}_+^{m \times n}$, $\mathbf{A} \in \mathbb{R}_+^{m \times r}$, $\mathbf{X} \in \mathbb{R}_+^{r \times n}$. Clearly, under the case of perfect decomposition, i.e., $\mathbf{Y}$ is equal to $\mathbf{AX}$, NMF model is equivalent to the noiseless BSS mixing model. This motivates researchers to exploit NMF schemes to solve the BSS problem [1].

In order to achieve NMF, several useful cost or measure functions have been proposed for particular applications. Let $\hat{\mathbf{Y}}$ be the decomposition of $\mathbf{Y}$. We list three major cost functions here.

- The first function is the Euclidean distance based function [8]

$$D(\mathbf{Y} \| \hat{\mathbf{Y}}) = \frac{1}{2} \| \mathbf{Y} - \hat{\mathbf{Y}} \|_2^2 = \frac{1}{2} \sum_{i=1}^{m} \sum_{j=1}^{n} (y_{ij} - \hat{y}_{ij})^2 \tag{2.22}$$

which measures the error of the given matrix and its decomposition. It is lower bounded by zero and vanishes if and only if $\mathbf{Y} = \hat{\mathbf{Y}}$.
- The second function is the Kullback-Leibler (KL) divergence based function [34]

$$D(\mathbf{Y} \| \hat{\mathbf{Y}}) = \sum_{i=1}^{m} \sum_{j=1}^{n} (y_{ij} \log \frac{y_{ij}}{\hat{y}_{ij}} - y_{ij} + \hat{y}_{ij}). \tag{2.23}$$

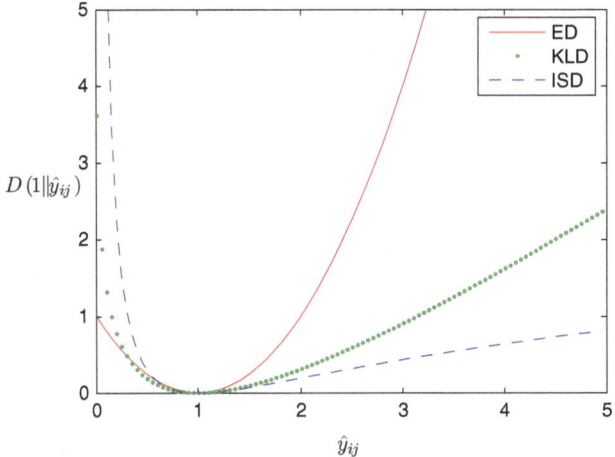

Fig. 2.9 Euclidean distance (ED), KL divergence (KLD) and IS divergence (ISD) versus $\hat{y}_{ij}$, where $y_{ij} = 1$ [35]

Similar to the Euclidean distance, this function is also lower bounded by zero and vanishes if and only if $\mathbf{Y} = \hat{\mathbf{Y}}$. Since it is not symmetric about $\mathbf{Y}$ and $\hat{\mathbf{Y}}$, it is called divergence.

- The third function is the Itakura-Saito (IS) divergence based function [35]

$$D(\mathbf{Y}\|\hat{\mathbf{Y}}) = \sum_{i=1}^{m}\sum_{j=1}^{n}(\frac{y_{ij}}{\hat{y}_{ij}} - \log\frac{y_{ij}}{\hat{y}_{ij}} - 1).\qquad(2.24)$$

Clearly, the IS divergence depends only on the ratio $\frac{y_{ij}}{\hat{y}_{ij}}$. This property is favorable when analyzing most audio signals such as music and speech, where the low frequency components have much higher energy than high frequency components.

Figure 2.9 shows the Euclidean distance, KL divergence and IS divergence under different $\hat{y}_{ij}$ varying from 0 to 5, where $y_{ij} = 1$. We can see that the KL and IS divergences are less sensitive to over-approximation than under-approximation.

There also exist some other divergence based cost functions, such as the α-divergence [36], β-divergence [37], $\alpha\beta$-divergence [38], and f-divergence [36]. Table 2.7 shows the afore-mentioned distance and divergence based cost functions, which are used for NMF.

Based on the Euclidean distance, the NMF optimization model to (2.21) is

$$\text{Minimize}: \ D = \frac{1}{2}\|\mathbf{Y} - \mathbf{AX}\|_2^2\qquad(2.25)$$

s.t. $\mathbf{A} \succeq \mathbf{0}$ and $\mathbf{X} \succeq \mathbf{0}$, where $\succeq$ denotes the component-wise inequality. Since NMF does not necessarily generate a desired result, one often needs to add some constraints

Table 2.7 Distance and divergence based cost functions

	Cost function $D(\mathbf{Y}\|\hat{\mathbf{Y}})$
Euclidean distance	$\frac{1}{2}\sum_{i=1}^{m}\sum_{j=1}^{n}(y_{ij}-\hat{y}_{ij})^2$
KL divergence	$\sum_{i=1}^{m}\sum_{j=1}^{n}(y_{ij}\log\frac{y_{ij}}{\hat{y}_{ij}}-y_{ij}+\hat{y}_{ij})$
IS divergence	$\sum_{i=1}^{m}\sum_{j=1}^{n}(\frac{y_{ij}}{\hat{y}_{ij}}-\log\frac{y_{ij}}{\hat{y}_{ij}}-1)$
β-divergence	$\sum_{i=1}^{m}\sum_{j=1}^{n}(\frac{y_{ij}^{\beta}}{\beta(\beta-1)}+\frac{\hat{y}_{ij}^{\beta}}{\beta}-\frac{y_{ij}\hat{y}_{ij}^{\beta-1}}{\beta-1}),\ \beta\in\mathbb{R},\ \beta\neq 0,1$
α-divergence	$\frac{1}{\alpha(1-\alpha)}\sum_{i=1}^{m}\sum_{j=1}^{n}\left(\alpha y_{ij}+(1-\alpha)\hat{y}_{ij}-y_{ij}^{\alpha}\hat{y}_{ij}^{1-\alpha}\right)$
$\alpha\beta$-divergence	$-\frac{1}{\alpha\beta}\sum_{i=1}^{m}\sum_{j=1}^{n}\left(y_{ij}^{\alpha}\hat{y}_{ij}^{\beta}-\frac{\alpha}{\alpha+\beta}y_{ij}^{\alpha+\beta}-\frac{\beta}{\alpha+\beta}\hat{y}_{ij}^{\alpha+\beta}\right),\ \alpha,\beta,\alpha+\beta\neq 0$
f-divergence	$\sum_{i=1}^{m}\sum_{j=1}^{n}\left(y_{ij}f(\frac{\hat{y}_{ij}}{y_{ij}})\right)$

(or regularization/penalty terms) into the model. A general constrained NMF model can be written as [39]

$$\text{Minimize}:\ D_J=\frac{1}{2}\|\mathbf{Y}-\mathbf{A}\mathbf{X}\|_2^2+\alpha J(\mathbf{A})+\beta J(\mathbf{X}) \qquad (2.26)$$
$$\text{s.t. } \mathbf{A}\succeq\mathbf{0}\text{ and }\mathbf{X}\succeq\mathbf{0}.$$

Dependent on the practical applications, different constraints could be considered. Some useful constraints are as follows.

- Volume based constraint on $\mathbf{A}$ [29]:

$$J(\mathbf{A})=\frac{1}{2(r-1)!}\det^2\left([\mathbf{1}\ \tilde{\mathbf{A}}^T]\right). \qquad (2.27)$$

Here, the matrix $\tilde{\mathbf{A}}\in\mathbb{R}^{(r-1)\times r}$ is calculated by

$$\tilde{\mathbf{A}}=\mathbf{U}^T(\mathbf{A}-\mu\mathbf{1}^T) \qquad (2.28)$$

where $\mathbf{U}\in\mathbb{R}^{m\times(r-1)}$ is formed by the $r-1$ most significant components of $\mathbf{Y}$ through principal component analysis and the column vector μ contains the means of the rows of $\mathbf{Y}$.

- Dispersion based constraint on $\mathbf{A}$ [40]:

$$J(\mathbf{A})=\text{Tr}\left(\mathbf{A}^T\mathbf{A}\right)-\frac{1}{m}\text{Tr}\left(\mathbf{A}^T\mathbf{E}\mathbf{A}\right) \qquad (2.29)$$

where $\mathrm{Tr}(\cdot)$ denotes the trace operator and $\mathbf{E}$ stands for the $m \times m$ matrix whose entries are all one.

- Temporal continuity based constraint on $\mathbf{X}$ [41]:

$$J(\mathbf{X}) = \sum_{j=1}^{r} \frac{1}{\sigma_j^2} \sum_{t=2}^{n} (x_{jt} - x_{j(t-1)})^2 \tag{2.30}$$

where $\sigma_j = \sqrt{(1/n) \sum_{t=1}^{n} x_{jt}^2}$ denotes the standard deviation of the jth component $\bar{\mathbf{x}}_j$ and $\bar{\mathbf{x}}_j$ is the jth row of $\mathbf{X}$.

- Dependence based constraint on $\mathbf{X}$ [42]:

$$J(\mathbf{X}) = \frac{1}{2} \left[\sum_{i=1}^{r} \log((\mathbf{X}\mathbf{X}^T)_{ii}) - \log(\det(\mathbf{X}\mathbf{X}^T)) \right]. \tag{2.31}$$

2.3.2 Estimation of Mixing Matrix and Source Signals

To solve the model (2.25), one can utilize the scheme based on the alternatively iterative multiplication updating rule in [34], together with the classic gradient based tool. From (2.25), the partial derivatives of the cost function D with respect to $\mathbf{X}$ and $\mathbf{A}$ are

$$\begin{cases} \frac{\partial D}{\partial \mathbf{X}} = \mathbf{A}^T \mathbf{A} \mathbf{X} - \mathbf{A}^T \mathbf{Y} \\ \frac{\partial D}{\partial \mathbf{A}} = \mathbf{A} \mathbf{X} \mathbf{X}^T - \mathbf{Y} \mathbf{X}^T \end{cases}. \tag{2.32}$$

According to the gradient based optimization rule, $\mathbf{X}$ and $\mathbf{A}$ can be updated by

$$\begin{cases} \mathbf{X} := \mathbf{X} - \eta_{\mathbf{X}} \otimes (\mathbf{A}^T \mathbf{A} \mathbf{X} - \mathbf{A}^T \mathbf{Y}) \\ \mathbf{A} := \mathbf{A} - \eta_{\mathbf{A}} \otimes (\mathbf{A} \mathbf{X} \mathbf{X}^T - \mathbf{Y} \mathbf{X}^T) \end{cases} \tag{2.33}$$

where $\otimes$ denotes the component-wise multiplication. To keep the nonnegativity of $\mathbf{X}$ and $\mathbf{A}$, the learning rates $\eta_{\mathbf{X}}$ and $\eta_{\mathbf{A}}$ are often chosen as

$$\begin{cases} \eta_{\mathbf{X}} = \frac{\mathbf{X}}{\mathbf{A}^T \mathbf{A} \mathbf{X}} \\ \eta_{\mathbf{A}} = \frac{\mathbf{A}}{\mathbf{A} \mathbf{X} \mathbf{X}^T} \end{cases}. \tag{2.34}$$

Then, the corresponding iteration formulae for $\mathbf{A}$ and $\mathbf{X}$ are as follows:

$$\begin{cases} \mathbf{X} := \mathbf{X} \otimes \frac{\mathbf{A}^T \mathbf{Y}}{\mathbf{A}^T \mathbf{A} \mathbf{X}} \\ \mathbf{A} := \mathbf{A} \otimes \frac{\mathbf{Y} \mathbf{X}^T}{\mathbf{A} \mathbf{X} \mathbf{X}^T} \end{cases}. \tag{2.35}$$

Table 2.8 NMF algorithm [34]

Step 1 (Initialization)	Randomly generate initial $\mathbf{A}$ and $\mathbf{X}$
Step 2 (Iteration)	While a stop criterion is not satisfied,
	(i) update $\mathbf{X}$ by
	$$\mathbf{X} := \mathbf{X} \otimes \frac{\mathbf{A}^T \mathbf{Y}}{\mathbf{A}^T \mathbf{A} \mathbf{X}};$$
	(ii) update $\mathbf{A}$ by
	$$\begin{cases} \mathbf{A} := \mathbf{A} \otimes \frac{\mathbf{Y} \mathbf{X}^T}{\mathbf{A} \mathbf{X} \mathbf{X}^T} \\ \mathbf{A} := \mathbf{A}(\text{diag}(\mathbf{1} \oslash (\sum_{i=1}^{m} \bar{\mathbf{a}}_i)^T)) \end{cases}$$
Step 3 (Estimation)	The final $\mathbf{A}$ and $\mathbf{X}$ are the estimates of the mixing matrix and the sources, respectively

Furthermore, to tackle the inevitable scaling issue, one often normalizes $\mathbf{A}$ by

$$\mathbf{A} := \mathbf{A}(\text{diag}(\mathbf{1} \oslash (\sum_{i=1}^{m} \bar{\mathbf{a}}_i)^T)) \qquad (2.36)$$

where $\oslash$ denotes the component-wise division, $\bar{\mathbf{a}}_i$ is the ith row of $\mathbf{A}$, and $\text{diag}(\mathbf{x})$ denotes a diagonal matrix whose diagonal entries correspond to the elements of the vector $\mathbf{x}$. The complete NMF algorithm is shown in Table 2.8.

Regarding the algorithms concerning the constrained NMF model (2.26), they are related to the exact constraints on the sources or the mixing matrix. For unmixing the hyper-spectral data, a source-constrained NMF method is proposed in [4]. Figure 2.10 shows the results of using this method and the traditional Kruse's method[2] to separate some hyper-spectral images. We can see that both methods obtain meaningful separation results.

In the following, we will further introduce some recently developed algorithms which apply volume constraint on $\mathbf{A}$, including both batch mode and online mode. According to the analysis in [32], a new volume constraint on $\mathbf{A}$ is $J_{\mathbf{A}} = \det(\mathbf{A}^T \mathbf{A})/2$. Then, (2.26) is simplified as

$$\text{Minimize}: \ D_J = \frac{1}{2}\|\mathbf{Y} - \mathbf{A}\mathbf{X}\|_2^2 + \frac{\alpha}{2}\det(\mathbf{A}^T \mathbf{A}) \qquad (2.37)$$

$$\text{s.t. } \mathbf{A} \succeq \mathbf{0} \text{ and } \mathbf{X} \succeq \mathbf{0}.$$

In this case, the derivatives of D_J with respect to $\mathbf{X}$ and $\mathbf{A}$ are

$$\begin{cases} \frac{\partial D_J}{\partial \mathbf{X}} = \mathbf{A}^T \mathbf{A} \mathbf{X} - \mathbf{A}^T \mathbf{Y} \\ \frac{\partial D_J}{\partial \mathbf{A}} = \mathbf{A} \mathbf{X} \mathbf{X}^T - \mathbf{Y} \mathbf{X}^T + \alpha(\det(\mathbf{A}^T \mathbf{A})\mathbf{A}(\mathbf{A}^T \mathbf{A})^{-1}) \end{cases} \qquad (2.38)$$

---

[2] Available: http://www.hgimaging.com/PDF/Kruse-JPL2002-AVIRIS-Hyperion.pdf.

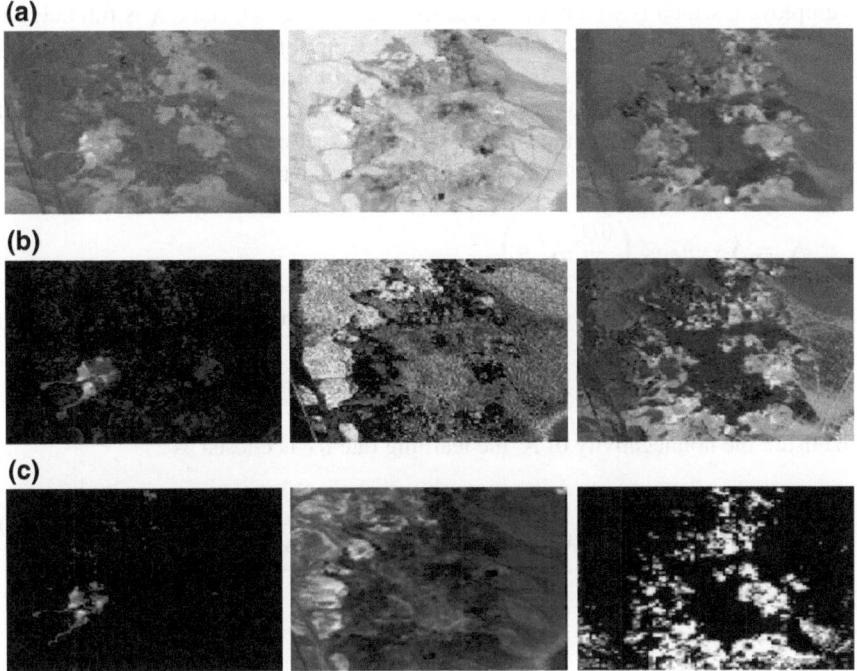

Fig. 2.10 Results of separating real-world hyper-spectral images by the constrained NMF algorithm and Kruse's algorithm. **a** Three source images; **b** Three recoveries using the constrained NMF; **c** Three recoveries using Kruse's method

A traditional gradient based method is utilized to update $\mathbf{X}$ as follows:

$$\begin{aligned} \mathbf{X} &:= \mathbf{X} - \eta_{\mathbf{X}} \otimes \frac{\partial D_J}{\partial \mathbf{X}} \\ &= \mathbf{X} - \eta_{\mathbf{X}} \otimes (\mathbf{A}^T \mathbf{A} \mathbf{X} - \mathbf{A}^T \mathbf{Y}) \\ &= \mathbf{X} - \eta_{\mathbf{X}} \otimes (\mathbf{A}^T \mathbf{A} \mathbf{X} + \delta_{\mathbf{X}} - \mathbf{A}^T \mathbf{Y} - \delta_{\mathbf{X}}). \end{aligned} \tag{2.39}$$

Here, $\delta_{\mathbf{X}}$ denotes a matrix with the same size of $\mathbf{X}$ and its entries all take the small positive value δ. It is used to avoid possible numerical instability. $\eta_{\mathbf{X}}$ denotes the learning rate. By setting $\eta_{\mathbf{X}} = \frac{\mathbf{X}}{\mathbf{A}^T \mathbf{A} \mathbf{X} + \delta_{\mathbf{X}}}$, then $\mathbf{X}$ can be updated by

$$\mathbf{X} := \mathbf{X} \otimes \frac{\mathbf{A}^T \mathbf{Y} + \delta_{\mathbf{X}}}{\mathbf{A}^T \mathbf{A} \mathbf{X} + \delta_{\mathbf{X}}}. \tag{2.40}$$

Now we consider the update of $\mathbf{A}$. As shown in (2.38), the partial derivative $\partial D_J / \partial \mathbf{A}$ includes the computation of the inverse matrix of $\mathbf{A}^T \mathbf{A}$. This may break the nonnegativity of $\mathbf{A}$. To conquer this obstacle, the so-called natural gradient (NG) is employed, which is widely discussed in [43, 44]. In fact, since $\mathbf{A}$ is full column rank, there exists a matrix $\mathbf{B}$ such that $\mathbf{BA} = \mathbf{I}_r$. Consequently, the parameter matrix $\mathbf{A}$ has a special algebraic structure, namely Li group structure, making the variables therein like a curved Riemann manifold. It is known that the natural gradient, instead of the ordinary gradient, is the steepest descent direction in the Riemann manifold [43, 44]. Hence, the natural gradient is utilized to update $\mathbf{A}$ as follows:

$$
\begin{aligned}
\mathbf{A} &:= \mathbf{A} - \eta_{\mathbf{A}} \otimes \left(\frac{\partial D_J}{\partial \mathbf{A}} \mathbf{A}^T \mathbf{A} \right) \\
&= \mathbf{A} - \eta_{\mathbf{A}} \otimes \left((\mathbf{AXX}^T - \mathbf{YX}^T + \alpha \det(\mathbf{A}^T\mathbf{A})\mathbf{A}(\mathbf{A}^T\mathbf{A})^{-1})\mathbf{A}^T\mathbf{A} \right) \quad (2.41) \\
&= \mathbf{A} - \eta_{\mathbf{A}} \otimes \left(\mathbf{AXX}^T\mathbf{A}^T\mathbf{A} + \alpha \det(\mathbf{A}^T\mathbf{A})\mathbf{A} + \delta_{\mathbf{A}} - \mathbf{YX}^T\mathbf{A}^T\mathbf{A} - \delta_{\mathbf{A}} \right).
\end{aligned}
$$

To ensure the nonnegativity of $\mathbf{A}$, the learning rate $\eta_{\mathbf{A}}$ is chosen as

$$
\eta_{\mathbf{A}} = \frac{\mathbf{A}}{\mathbf{AXX}^T\mathbf{A}^T\mathbf{A} + \alpha \det(\mathbf{A}^T\mathbf{A})\mathbf{A} + \delta_{\mathbf{A}}}. \quad (2.42)
$$

Thus, $\mathbf{A}$ can be updated by

$$
\mathbf{A} := \mathbf{A} \otimes \frac{\mathbf{YX}^T\mathbf{A}^T\mathbf{A} + \delta_{\mathbf{A}}}{\mathbf{AXX}^T\mathbf{A}^T\mathbf{A} + \alpha \det(\mathbf{A}^T\mathbf{A})\mathbf{A} + \delta_{\mathbf{A}}}. \quad (2.43)
$$

The NG based minimum volume constrained NMF (NG-MVC-NMF) algorithm [32] is summarized in Table 2.9.

In addition to the batch algorithm for the volume based NMF, the corresponding online learning version has also been developed. Compared with the batch mode which usually suffers from large storage requirement and high computational complexity when the observations are large scale, the online mode or incremental learning scheme is particularly appealing owing to its low computational cost. Table 2.10 shows the incremental NMF with volume constraint (INMF-VC) [30].

Table 2.9 NG-MVC-NMF algorithm [32]

Step 1 (Initialization)	Randomly generate initial $\mathbf{A}$ and $\mathbf{X}$. Set $\alpha > 0$ and $\delta = 10^{-6}$
Step 2 (Iteration)	While a stop criterion is not satisfied,
	(i) update $\mathbf{X}$ by $$\mathbf{X} := \mathbf{X} \otimes \frac{\mathbf{A}^T\mathbf{Y}+\delta_{\mathbf{X}}}{\mathbf{A}^T\mathbf{AX}+\delta_{\mathbf{X}}};$$
	(ii) update $\mathbf{A}$ by $$\begin{cases} \mathbf{A} := \mathbf{A} \otimes \frac{\mathbf{YX}^T\mathbf{A}^T\mathbf{A}+\delta_{\mathbf{A}}}{\mathbf{AXX}^T\mathbf{A}^T\mathbf{A}+\alpha\det(\mathbf{A}^T\mathbf{A})\mathbf{A}+\delta_{\mathbf{A}}} \\ \mathbf{A} := \mathbf{A}(\mathrm{diag}(1 \oslash (\tilde{\mathbf{a}}))) \end{cases}.$$
Step 3 (Estimation)	The final $\mathbf{A}$ and $\mathbf{X}$ are the estimates of the mixing matrix and the sources, respectively

Table 2.10 INMF-VC algorithm [30]

Step 1 (Initialization)	Set an initial sample number p for learning process. Then, (i) project the collected $k = p$ samples $\mathbf{y}_1, \ldots, \mathbf{y}_k$ to be $[\tilde{\mathbf{y}}_1, \ldots, \tilde{\mathbf{y}}_k]$ on the hyperplane $\Pi : \sum_{i=1}^{m} \tilde{y}_i = 1$, and construct the initial matrix $\tilde{\mathbf{Y}}_k = [\tilde{\mathbf{y}}_1, \ldots, \tilde{\mathbf{y}}_k]$; (ii) obtain $\mathbf{A}_k$ and $\tilde{\mathbf{X}}_k$ from $\tilde{\mathbf{Y}}_k$ by using the normal NMF algorithm
Step 2 (Learning)	For the $(k+1)$th sample $\mathbf{y}_{k+1}$, (i) project it to be $\tilde{\mathbf{y}}_{k+1}$ by $[\tilde{\mathbf{y}}_{k+1}]_i = [\mathbf{y}_{k+1}]_i / \sum_{j=1}^{m} [\mathbf{y}_{k+1}]_j$; (ii) let $\mathbf{A}_{k+1} = \mathbf{A}_k$ and update $\tilde{\mathbf{x}}_{k+1}$ by $$\tilde{\mathbf{x}}_{k+1} := \tilde{\mathbf{x}}_{k+1} \otimes \frac{\mathbf{A}_{k+1}^T \tilde{\mathbf{h}}_{k+1} + \delta_{\tilde{\mathbf{x}}_{k+1}}}{\mathbf{A}_{k+1}^T \mathbf{A}_{k+1} \tilde{\mathbf{x}}_{k+1} + \delta_{\tilde{\mathbf{x}}_{k+1}}};$$ (iii) update $\mathbf{A}_{k+1}$ by $$\mathbf{A}_{k+1} := \mathbf{A}_{k+1} \otimes \frac{\left(\alpha \tilde{\mathbf{Y}}_k \tilde{\mathbf{X}}_k^T + \beta \tilde{\mathbf{y}}_{k+1} \tilde{\mathbf{x}}_{k+1}^T \right) \mathbf{A}_{k+1}^T \mathbf{A}_{k+1} + \delta_{\mathbf{A}_{k+1}}}{\left(\alpha \mathbf{A}_{k+1} \tilde{\mathbf{X}}_k \tilde{\mathbf{X}}_k^T + \beta \mathbf{A}_{k+1} \tilde{\mathbf{x}}_{k+1} \tilde{\mathbf{x}}_{k+1}^T \right) \mathbf{A}_{k+1}^T \mathbf{A}_{k+1} + \beta \mu \mathbf{A}_{k+1} + \delta_{\mathbf{A}_{k+1}}}$$
Step 3 (Estimation)	Estimate the $(k+1)$th column of $\mathbf{X}$ by $\mathbf{x}_{k+1} = \tilde{\mathbf{x}}_{k+1} / \sum_{j=1}^{m} [\mathbf{y}_{k+1}]_j$. If $k+1$ is less than the number of the total samples, let $k = k+1$ and go to Step 2

Table 2.11 NGMCA algorithm [45]

Step 1 (Initialization)	Set a maximum iteration number K and the initial values for $\mathbf{A}_0$, $\mathbf{X}_0$ and λ_1
Step 2 (Iteration)	For $k = 1, 2, \ldots, K$ (i) normalize the columns of $\mathbf{A}_{k-1}$; (ii) update $\mathbf{X}_k$ by $\mathbf{X}_k = \mathrm{argmin}_{\mathbf{X} \geq 0} \frac{1}{2} \|\mathbf{Y} - \mathbf{A}_{k-1}\mathbf{X}\|_2^2 + \lambda_k \|\mathbf{X}\|_1$; (iii) update $\mathbf{A}_k$ by $\mathbf{A}_k = \mathrm{argmin}_{\mathbf{A} \geq 0} \frac{1}{2} \|\mathbf{Y} - \mathbf{A}\mathbf{X}_k\|_2^2$; (iv) Select $\lambda_{k+1} \leq \lambda_k$
Step 3 (Estimation)	$\mathbf{A}_K$ and $\mathbf{X}_K$ are the estimates of the mixing matrix and the source matrix, respectively

Furthermore, the nonnegative generalized morphological component analysis (NGMCA) method [45] is proposed to deal with DCA in the situations where the measurements or the observations are polluted by noise. Different from the traditional NMF algorithms which are usually applicable to determined mixing systems, NGMCA can also be applied to underdetermined mixing systems. Table 2.11 shows the NGMCA algorithm with soft threshold.

2.3.3 Algorithm Analysis

As for the initialization of the decomposition process, there are many useful methods which are verified to be efficient. The singular value decomposition based scheme is one of the best methods for initialization, which has advantages in approximating the data matrix [46]. Other initialization methods include spherical k-means clustering [47], principal component analysis, fuzzy clustering and Gabor wavelets [48], etc.

Regarding the convergence of algorithms, Lin gives a detailed convergence analysis of the multiplicative update algorithms which are widely used in NMF based methods [49], Yang and Yi propose a novel scheme to analyze the convergence of the constrained NMF with application to BSS [50], and Badeau et al. analyze the stability of these algorithms [51].

With regard to the uniqueness of NMF, it is analyzed under different conditions in [52]. In [53], Donoho and Stodden show that the NMF is unique if the involved data satisfies three rules: (a) generative model, (b) separability, and (c) complete factorial sampling. In [54], Laurberg et al. propose the following theorem.

Theorem 2.9 *If* $rank(\mathbf{Y}) = r$, *the NMF* $\mathbf{Y} = \mathbf{AX}$ *is unique if and only if the nonnegative orthant is the only simplicial cone* $\mho$ *with r extreme rays that satisfies*

$$cone(\mathbf{A}^T) \subseteq \mho \subseteq dcone(\mathbf{X}) \tag{2.44}$$

where $dcone(\mathbf{X})$ *denotes the dual cone of* $cone(\mathbf{X})$.

Furthermore, Schachtner et al. propose a determinant criterion to constrain the solutions of NMF problems and achieve unique and optimal solutions in a general setting [55].

References

1. A. Cichocki, R. Zdunek, A.H. Phan, S. Amari, *Non-Negative Matrix and Tensor Factorization: Applications to Exploratory Multi-Way Data Analysis and Blind Source Separation* (Wiley-Blackwell, Oxford, 2009)
2. P. Sajda, S. Du, T. Brown, R. Stoyanova, D. Shungu, L.P.X. Mao, Nonnegative matrix factorization for rapid recovery of constituent spectra in magnetic resonance chemical shift imaging of the brain. IEEE Trans. Med. Imaging **23**(12), 1453–1465 (2004)
3. R.H.C. Gobinet, E. Perrin, Application of non-negative matrix factorization to fluorescence spectroscopy, in: *Proceedings 12th European Signal Processing Conference*, pp. 1095–1098 (2004)
4. Z. Yang, G. Zhou, S. Xie, S. Ding, J. Yang, J. Zhang, Blind spectral unmixing based on sparse nonnegative matrix factorization. IEEE Trans. Image Process. **20**(4), 1112–1125 (2011)
5. T.H. Chan, W.K. Ma, C.Y. Chi, Y. Wang, A convex analysis framework for blind separation of non-negative sources. IEEE Trans. Signal Process. **56**(10), 5120–5134 (2008)
6. T.H. Chan, C.Y. Chi, Y.M. Huang, W.K. Ma, A convex analysis-based minimum-volume enclosing simplex algorithm for hyperspectral unmixing. IEEE Trans. Signal Process. **57**(11), 4418–4432 (2009)

7. W.S.B. Ouedraogo, A. Souloumiac, M. Jaïdane, C. Jutten, Non-negative blind source separation algorithm based on minimum aperture simplicial cone. IEEE Trans. Signal Process. **62**(2), 376–389 (2014)

8. D.D. Lee, H.S. Seung, Learning of the parts of objects by non-negative matrix factorization. Nature **401**(6755), 788–791 (1999)

9. D. Donoho, M. Elad, V. Temlyakov, Stable recovery of sparse overcomplete representations in the presence of noise. IEEE Trans. Inf. Theory **52**(1), 6–18 (2006)

10. R. Chartrand, Exact reconstruction of sparse signals via nonconvex minimization. IEEE Signal Process. Lett. **14**(10), 707–710 (2007)

11. Y. Li, A. Cichocki, S. Amari, Analysis of sparse representation and blind source separation. Neural Comput. **16**(6), 1193–1234 (2004)

12. P.O. Hoyer, Non-negative matrix factorization with sparseness constraints. J. Mach. Learn. Res. **5**, 1457–1469 (2004)

13. Z. Yang, Y. Xiang, S. Xie, S. Ding, Y. Rong, Nonnegative blind source separation by sparse component analysis based on determinant measure. IEEE Trans. Neural Netw. Learn. Syst. **23**(10), 1601–1610 (2012)

14. G. Rath, C. Guillemot, J. Fuchs, Sparse approximations for joint source-channel coding, in Proc. IEEE 10th Workshop on Multimedia Signal Processing, 2008, pp. 481–485.

15. J. Karvanen, A. Cichocki, Measuring sparseness of noisy signals, in: *Proceedings Fourth International Symposium on Independent Component Analysis and Blind Signal Separation*, pp. 125–130 (2003)

16. S. Rickard, M. Fallon, The Gini index of speech, in: 38th *Proc. Conf. Inf. Sci. Syst*, Princeton, NJ, (2004)

17. B.A. Olshausen, D.J. Field, Sparse coding of sensory inputs. Curr. Opin. Neurobiol. **14**(4), 481–487 (2004)

18. B. Rao, K. Kreutz-Delgado, An affine scaling methodology for best basis selection. IEEE Trans. Signal Process. **47**(1), 187–200 (1999)

19. A. Bronstein, M. Bronstein, M. Zibulevsky, Y.Y. Zeevi, Sparse ICA for blind separation of transmitted and reflected images. Int. J. Imaging Syst. Technol. **15**(1), 84–91 (2005)

20. N. Hurley, S. Rickard, Comparing measures of sparsity. IEEE Trans. Inf. Theory **55**(10), 4723–4741 (2009)

21. P. Georgiev, F. Theis, A. Cichocki, Sparse component analysis and blind source separation of underdetermined mixtures. IEEE Trans. Neural Netw. **16**(4), 992–996 (2005)

22. A.M. Bruckstein, M. Elad, M. Zibulevsky, On the uniqueness of nonnegative sparse solutions to underdetermined systems of equations. IEEE Trans. Inf. Theory **54**(11), 4813–4820 (2008)

23. E. Candès, T. Tao, Decoding by linear programming. IEEE Trans. Inf. Theory **51**(12), 4203–4215 (2005)

24. D. Donoho, J. Tanner, Sparse nonnegative solution of underdetermined linear equations by linear programming. PNAS **102**(27), 9446–9451 (2005)

25. F.Y. Wang, C.Y. Chi, T.H. Chan, Y. Wang, Nonnegative least-correlated component analysis for separation of dependent sources by volume maximization. IEEE Trans. Pattern Anal. Mach. Intell. **32**(5), 875–888 (2010)

26. J.M.P. Nascimento, J.M.B. Dias, Vertex component analysis: A fast algorithm to unmix hyperspectral data. IEEE Trans. Geosci. Remote Sens. **43**(4), 898–910 (2005)

27. Y. Wang, J. Xuan, R. Srikanchana, P.L. Choyke, Modeling and reconstruction of mixed functional and molecular patterns. Int. J. Biomed. Imaging **2006**, 1–9 (2006)

28. Z. Yang, Y. Xiang, Y. Rong, S. Xie, Projection-pursuit-based method for blind separation of nonnegative sources. IEEE Trans. Neural Netw. Learn. Syst. **24**(1), 47–57 (2013)

29. L. Miao, H. Qi, Endmember extraction from highly mixed data using minimum volume constrained nonnegative matrix factorization. IEEE Trans. Geosci. Remote Sens **45**(3), 765–777 (2007)

30. G. Zhou, Z. Yang, S. Xie, J. Yang, Online blind source separation using incremental nonnegative matrix factorization with volume constraint. IEEE Trans. Neural Netw. **22**(4), 550–560 (2011)

31. S. Lopez, P. Horstrand, G.M. Callico, J.F. Lopez, R. Sarmiento, A novel architecture for hyper-spectral endmember extraction by means of the modified vertex component analysis (MVCA) algorithm. IEEE. J. Sel. Top. Appl. Earth Obs. Remote Sens. **5**(6), 1837–1848 (2012)
32. G. Zhou, S. Xie, Z. Yang, J. Yang, Z. He, Minimum-volume-constrained nonnegative matrix factorization: enhanced ability of learning parts. IEEE Trans. Neural Netw. **22**(10), 1626–1637 (2011)
33. A. Cichocki, R. Zdunek, S.I. Amari, New algorithms for nonnegative matrix factorization in applications to blind source separation, in: *Proceedings 2006 IEEE International Conference on Acoustics, Speech and Signal Processing*, pp. 5479–5482 (2006)
34. D.D. Lee, H.S. Seung, Algorithms for non-negative matrix factorization. Adv. Neural Inf. Process. Syst. **13**, 556–562 (2001)
35. H. Sawada, H. Kameoka, S. Araki, N. Ueda, Multichannel extensions of non-negative matrix factorization with complex-valued data. IEEE Trans. Audio, Speech, Lang. Process. **21**(5), 971–982 (2013)
36. A. Cichocki, H. Lee, Y.D. Kim, S. Choi, Non-negative matrix factorization with α-divergence. Pattern Recognit. Lett. **29**(9), 1433–1440 (2008)
37. V.Y.F. Tan, C. Févotte, Automatic relevance determination in nonnegative matrix factorization with the β-divergence. IEEE Trans. Pattern Anal. Mach. Intell. **35**(7), 1592–1605 (2013)
38. A. Cichocki, S. Cruces, S. Amari, Generalized alpha-beta divergences and their application to robust nonnegative matrix factorization. Entropy **13**, 134–170 (2011)
39. V.P. Pauca, J. Piper, R.J. Plemmons, Nonnegative matrix factorization for spectral data analysis. Linear Algebra Appl. **416**(1), 29–47 (2006)
40. A. Huck, M. Guillaume, J. Blanc-Talon, Minimum dispersion constrained nonnegative matrix factorization to unmix hyperspectral data. IEEE Trans. Geosci. Remote Sens. **48**(6), 2590–2612 (2010)
41. T. Virtanen, Monaural sound source separation by nonnegative matrix factorization with temporal continuity and sparseness criteria. IEEE Trans. Audio, Speech, Lang. Process. **15**(3), 1066–1074 (2007)
42. Y. Zhang, Y. Fang, A NMF algorithm for blind separation of uncorrelated signals, in: *Proceedings 2007 IEEE International Conference on Wavelet Analysis and Pattern Recognition*, November 2–4, Beijing, China, pp. 999–1003 (2007)
43. S. Amari, Natural gradient works efficiently in learning. Neural Comput. **10**(2), 251–276 (1998)
44. A. Cichocki, S. Amari, *Adaptive Blind Signal and Image Processing: Learning Algorithms and Applications* (Wiley, New York, 2003)
45. J. Rapin, J. Bobin, A. Larue, J.-L. Starck, Sparse and non-negative BSS for noisy data. IEEE Trans. Signal Process. **61**(22), 5620–5632 (2013)
46. C. Boutsidis, E. Gallopoulos, SVD based initialization: a head start for nonnegative matrix factorization. Pattern Recogn. **41**(4), 1350–1362 (2008)
47. S. Wild, J. Curry, A. Dougherty, Improving non-negative matrix factorizations through structured initialization. Pattern Recogn. **37**(11), 2217–2232 (2004)
48. Z. Zheng, J. Yang, Y. Zhu, Initialization enhancer for non-negative matrix factorization. Eng. Appl. Artif. Intell. **20**(1), 101–110 (2007)
49. C.-J. Lin, On the convergence of multiplicative update algorithms for nonnegative matrix factorization. IEEE Trans. Neural Netw. **18**(6), 1589–1596 (2007)
50. S. Yang, Z. Yi, Convergence analysis of non-negative matrix factorization for BSS algorithm. Neural Process. Lett. **31**(1), 45–64 (2010)
51. R. Badeau, N. Bertin, E. Vincent, Stability analysis of multiplicative update algorithms and application to nonnegative matrix factorization. IEEE Trans. Neural Netw. **21**(12), 1869–1881 (2010)
52. K. Huang, N.D. Sidiropoulos, A. Swami, Non-negative matrix factorization revisited: uniqueness and algorithm for symmetric decomposition. IEEE Trans. Signal Process. **62**(1), 211–224 (2014)
53. D.L. Donoho, V.C. Stodden, When does non-negative matrix factorization give a correct decomposition into parts? Adv. Neural Inf. Process. Syst. **16**, 1141–1148 (2003)

54. H. Laurberg, M.G. Christensen, M.D. Plumbley, L.K. Hansen, S.H. Jensen, Theorems on positive data: on the uniqueness of NMF. Comput. Intell. Neurosci. **2008**, 1–9 (2008)
55. R. Schachtner, G. Pöppel, E.W. Lang, Towards unique solutions of non-negative matrix factorization problems by a determinant criterion. Digit. Signal Process. **21**(4), 528–534 (2011)

Chapter 3
Dependent Component Analysis Using Time-Frequency Analysis

Abstract Sparsity is an important property shared by many kinds of signals in numerous practical applications. These signals are sparse to some extent in different representation domains, such as time domain, frequency domain or time-frequency domain. In recent years, sparsity has been widely exploited to solve the problem of underdetermined blind source separation (UBSS), where the number of sources exceeds that of the observed mixtures. In fact, the sparsity assumption can also be satisfied by some dependent source signals. For these signals, it is possible to find a number of areas in some representation domains, where the source signals are not active, that is, signals are sparse in theses areas. The sparsity property provides a possibility for the blind separation of dependent sources. In this chapter, the sparsity of dependent sources in the time-frequency (TF) domain will be exploited to achieve blind source separation, where time-frequency analysis (TFA) will be used as a powerful tool for dependent component analysis (DCA). We will also show that for those non-sparse signals whose auto-source points and cross-source points do not overlap in the TF plane, they can be separated by using TFA if the underdetermined mixing system is known.

Keywords Time-frequency analysis · Spatial time-frequency distribution · Quadratic TFD

3.1 Fundamentals of TFA

Conventional approaches for blind source separation (BSS) usually depend on an ideal assumption that the mixed source signals are statistically uncorrelated or independent. However, there are lots of practical applications, where the uncorrelation or independence assumption cannot be satisfied. Therefore, we need to exploit other properties of source signals to achieve blind separation of dependent sources. Sparsity has been widely used to solve the problem of UBSS [1–15]. Since the mixing matrix is column rank deficient in the underdetermined case, it is a very challenging task to blindly recover the source signals from their mixtures, even if the mixing matrix is known. Usually, signals have sparser representations in TF domain than in time domain, so it is more preferable to make use of the TF representations of signals in the development of sparsity-based UBSS algorithms. In addition to sparsity, other TF property of sources could also be employed to perform UBSS [16]. In this chapter, we will discuss how to use the TFA tool to achieve dependent component analysis [17–40]. We would like to note here that different from the nonnegativity based methods for DCA, the TFA-based methods do not require the source signals or the mixing matrix to be nonnegative.

Before introducing the related approaches, we first review some fundamentals of time-frequency analysis or TFA. Time-frequency analysis is a powerful tool for analyzing nonstationary signals, whose frequency contents vary in time [18]. Since time-frequency representation provides a distribution of signal energy versus time and frequency simultaneously, it is also called as time-frequency distribution (TFD) [18]. Quadratic TFD is one of the most widely-used time-frequency distributions. Although it suffers from the so-called "cross-terms" problem, it can provide higher TF resolution than linear TFD, such as short-time Fourier transform (STFT) [18]. Based on the above considerations, this chapter will mainly employ a typical kind of Quadratic TFDs, i.e., Cohen's class of TFDs, to obtain the TF representations of source signals. The discrete-time form of Cohen's class of TF representations, for signal $x_i(t)$, has the following definition [17]:

$$D_{x_i x_i}(t, f) = \sum_{h=-\infty}^{\infty} \sum_{k=-\infty}^{\infty} \phi(k, h) x_i(t + k + h)$$
$$\times x_i(t + k - h) e^{-j4\pi f h} \tag{3.1}$$

where $\phi(k, h)$ is the kernel function of both the time and lag variables. The cross-TFD of two signals $x_i(t)$ and $x_j(t)$ is given by [19]

$$D_{x_i x_j}(t, f) = \sum_{h=-\infty}^{\infty} \sum_{k=-\infty}^{\infty} \phi_{i,j}(k, h) x_i(t + k + h)$$
$$\times x_j(t + k - h) e^{-j4\pi f h} \tag{3.2}$$

where the entry $\phi_{i,j}(k, h)$ is the kernel function associated with the signals $x_i(t)$ and $x_j(t)$.

Next, let us review the existing works about BSS based on the sparsity assumption. In many early works, the sparsity of source signals in time domain is exploited to develop BSS algorithms. Since time-frequency domain can provided sparser representations for source signals than time domain, more research efforts have been devoted to making use of the time-frequency representations of signals to develop blind separation approaches. Recently, the combination of the sparsity and the spatial diversity of signals leads to the spatial time-frequency distribution (STFD) of signals [19], where the STFD matrix of a signal vector $\mathbf{x}(t) = [x_1(t), x_2(t), \ldots, x_r(t)]^T$ is defined as follows [19]

$$\mathbf{D_{xx}}(t, f) = \begin{bmatrix} D_{x_1x_1}(t, f) & \cdots & D_{x_1x_r}(t, f) \\ \vdots & \ddots & \vdots \\ D_{x_rx_1}(t, f) & \cdots & D_{x_rx_r}(t, f) \end{bmatrix}. \tag{3.3}$$

It is shown by Belouchrani et al. [19] that a BSS algorithm utilizing the above STFDs of source signals possesses two obvious advantages. Firstly, it can separate Gaussian sources with identical spectral shapes but with different time-frequency localization properties. Secondly, the effect of spreading the noise power while localizing the source energy in the time-frequency domain increases the robustness of the algorithm with respect to noise.

Since the pioneer works of Belouchrani et al., a number of algorithms using spatial time-frequency distribution have been proposed to achieve BSS [9, 12–14, 18]. These methods exploit different sparsity assumptions about the TFDs of sources. Specifically, [13, 14] propose the cluster-based algorithms, which are based on the TF-disjoint assumption, i.e., there exists at most one active source at any TF point. Clearly, this is a rather restrictive constraint about the source signals. When this condition cannot be satisfied, the separation performance will degrade greatly at those overlapped TF points [18].

In order to overcome this drawback and relax the related restrictions, some subspace-based algorithms have been developed [9, 18]. In these algorithms, the sparsity assumptions about the source signals are relaxed to some extent by exploiting the spatial structures of the time-frequency representations of the source signals and mixed signals. In [9, 18], the TF distributions of different sources are allowed to overlap to the extent that the number of active sources at any TF point is strictly less than that of the mixed signals.

Although the assumption used in [9, 18] is more relaxing than the TF-disjoint assumption, it may be still too strict for some applications. For instance, in an application environment with two sensors, the above sparsity condition will reduce to the TF-disjoint condition, and require that at most one source signal is active at any time-frequency point. Consequently, the subspace-based approaches in [9, 18] cannot effectively separate any overlapped sources in this case. Recently, the sparsity conditions have been further relaxed in [7], such that the number of active sources at each TF point is permitted to be equal to or less than that of the mixtures. The method in [29] requires a source sparsity condition which is even weaker than that of [7] but it restricts the number of sources to be less than twice of that of the mixtures.

Source sparsity condition is exempted from the TFA-based source recovery method in [16], i.e., the source signals can be non-sparse. Instead, this method requires that the auto-source points and cross-source points of the source signals do not overlap in the TF plane. Moreover, it restrict the number of mixtures to be no less than three, and the number of sources to be less than twice of that of the mixtures.

We would like to note that the UBSS approaches based on TFA are usually composed of two stages. In the first stage, the mixing matrix is estimated. After that, the source signals are recovered by exploiting the estimated mixing matrix. In the remainder of this chapter, the TFA-based methods for mixing matrix estimation and source recovery will be introduced, respectively.

3.2 TFA-Based Methods for Mixing Matrix Estimation

3.2.1 System Model and Existing Works

Let us consider the $m \times r$ instantaneous mixing system in (1.7): $\mathbf{y}(t) = \mathbf{A}\mathbf{x}(t)$, where $\mathbf{x}(t) = [x_1(t), x_2(t), \ldots, x_r(t)]^T$ is the r-dimensional source signal vector, $\mathbf{y}(t) = [y_1(t), y_2(t), \ldots, y_m(t)]^T$ is the m-dimensional mixed signal vector, and $\mathbf{A} = [\mathbf{a}_1, \mathbf{a}_2, \ldots, \mathbf{a}_r]$ is the $m \times r$ mixing matrix. In the mixing model, t is an index given to each sample of signals. As a generalized index, t has different meanings in different domains. For example, t stands for a time instant in the time domain, while in the TF domain, the index t denotes a sample point in the TF plain. The column vector $\mathbf{a}_i$ in $\mathbf{A}$ is called as the steering vector corresponding to the ith source signal $x_i(t)$, where $i = 1, 2, \ldots, r$ [7]. In order to achieve blind separation for dependent sources, one can exploit another important property of source signals, i.e. sparsity. Sparse representation provides a good way for solving the DCA problem [5, 8, 36].

Estimating the mixing matrix $\mathbf{A}$ is usually the first stage of the sparsity based UBSS approaches. So far, a number of blind approaches have been developed to identify the mixing matrix by exploiting the sparsity property of source signals. By assuming that at most one source signal possesses the dominant energy at each sample point, which is called as the W-disjoint orthogonal condition (or approximate W-disjoint orthogonality), early works [9, 10] use the clustering approaches to estimate the steering vectors in the mixing matrix $\mathbf{A}$, and then obtain the estimation for $\mathbf{A}$. For example, a so-called DUET algorithm is proposed in [10, 11] to estimate the mixing matrix by exploiting the ratios of the time-frequency transforms of the observed mixtures under the above W-disjoint orthogonal condition. In [32–34], the sparsity constraint about the source signals is relaxed to some extent by allowing the energy distributions of the sources to overlap. Specifically, it is not required in [32–34] that each sample point should have at most one active source. Instead, a weaker sparsity condition is needed, i.e., there are some adjacent sample regions where only one source is active. Based on this sparsity constraint, the so-called TIFROM algorithm is proposed to tackle the BSS problem in [32–34]. More recently, as an extension of the DUET algorithm and the TIFROM algorithm, an effective UBSS algorithm is developed in [5] to estimate the mixing matrix under a milder sparsity condition, i.e., there exist some isolated (or discrete) sample points at which only one single source is active. Clearly, it is weaker than the sparsity constraints required by the DUET algorithm and the TIFROM algorithm. The algorithm in [5] firstly identifies and collects those sample points with only one active source, and then achieve the estimation for the mixing matrix by grouping the output signal vectors associated with the same source signal.

3.2.2 *Mixing Matrix Estimation Under Relaxed TF Sparsity*

It is notable that all of the blind identification approaches mentioned in the previous subsection need some sample points to possess at most one active source signal. That is, there must exist some sample points where the energy distributions of the source signals do not overlap. Such a sparsity restriction could be too difficult to satisfy in some practical applications, where all sample points have more than one active source. Interestingly, this sparsity restriction is successfully removed in [8] at the cost of an additional condition about the richness of the source signals. Specifically, [8] allows the distributions of source signals to overlap at all sample points if the following richness constraint about the source signals holds:

(A3.1) The source signals are sufficiently rich in the sense that for any $m - 1$ sources in all r sources, there must exist at least m sample points at which these $m - 1$ sources are active.

In addition, the algorithm in [8] also requires, respectively, the following conditions on the mixing matrix and the sparsity of the sources:

(A3.2) Any m column vectors in the mixing matrix $\mathbf{A}$ are linearly independent.
(A3.3) At most $m - 1$ sources among the r sources are active at any sample point.

In [8], the basic idea concerning the estimation of the mixing matrix is based on the following three important observations:

- According to the assumptions (A3.3), all observed mixtures $\mathbf{y}(t)$ must lie in one $(m - 1)$-dimensional subspace.
- From the assumptions (A3.1) and (A3.2), it holds that the number of the above subspaces is equal to C_r^{m-1}, each of which is spanned by $m - 1$ steering vectors in the mixing matrix $\mathbf{A}$.
- Any steering vector $\mathbf{a}_i (i = 1, 2, \ldots, r)$ in the mixing matrix $\mathbf{A}$ must be the intersection of C_{r-1}^{m-2} subspaces, which means that if all C_r^{m-1} subspaces can be found, their intersection can be viewed as the estimate of the steering vector.

Based on these essential observations, an algorithm is proposed in [8] to blindly identify the mixing matrix, which is shown in Table 3.1.

Table 3.1 Mixing matrix identification algorithm in [8]

Step 1	Obtain the estimates of the C_r^{m-1} subspaces by subspace clustering all observed mixtures $\mathbf{y}(t)$
Step 2	Compute the normal vectors to those subspaces obtained in Step 1
Step 3	Cluster these normal vectors into r subspaces $\mathscr{S}_i (i = 1, 2, \ldots, r)$, each of which contains C_{r-1}^{m-2} normal vectors
Step 4	Estimate all steering vectors $\mathbf{a}_i (i = 1, 2, \ldots, r)$ in the mixing matrix $\mathbf{A}$ as the normal vectors of these r subspaces $\mathscr{S}_i (i = 1, 2, \ldots, r)$

From the above analysis, we can easily find that the approach in [8] does not require the source signals to be mutually uncorrelated or independent. In other words, this approach can also be used to conduct depend component analysis. However, [8] depends on an assumption about the richness of the source signals, i.e., the assumption (A3.1). It is worth mentioning that this assumption is still strict and difficult to satisfy in many practical applications. In order to illustrate this point, we consider a simple blind source separation scenario with $m = 3$ mixtures and $r = 4$ source signals $x_1(t), x_2(t), x_3(t)$ and $x_4(t)$, which means that the mixing matrix $\mathbf{A}$ is of dimension 3×4. According to the assumption (A3.1), these 4 source signals must satisfy the following $C_r^{m-1} = C_4^2 = 6$ conditions:

(c1) There are at least $m = 3$ time instants at which $x_1(t)$ and $x_2(t)$ are active.
(c2) There are at least $m = 3$ time instants at which $x_1(t)$ and $x_3(t)$ are active.
(c3) There are at least $m = 3$ time instants at which $x_1(t)$ and $x_4(t)$ are active.
(c4) There are at least $m = 3$ time instants at which $x_2(t)$ and $x_3(t)$ are active.
(c5) There are at least $m = 3$ time instants at which $x_2(t)$ and $x_4(t)$ are active.
(c6) There are at least $m = 3$ time instants at which $x_3(t)$ and $x_4(t)$ are active.

Regarding the above six conditions, let us consider two sets of source signals $\{x_1(t), x_2(t), x_3(t), x_4(t)\}$ and $\{s_1(t), s_2(t), s_3(t), s_4(t)\}$, which are shown in the left

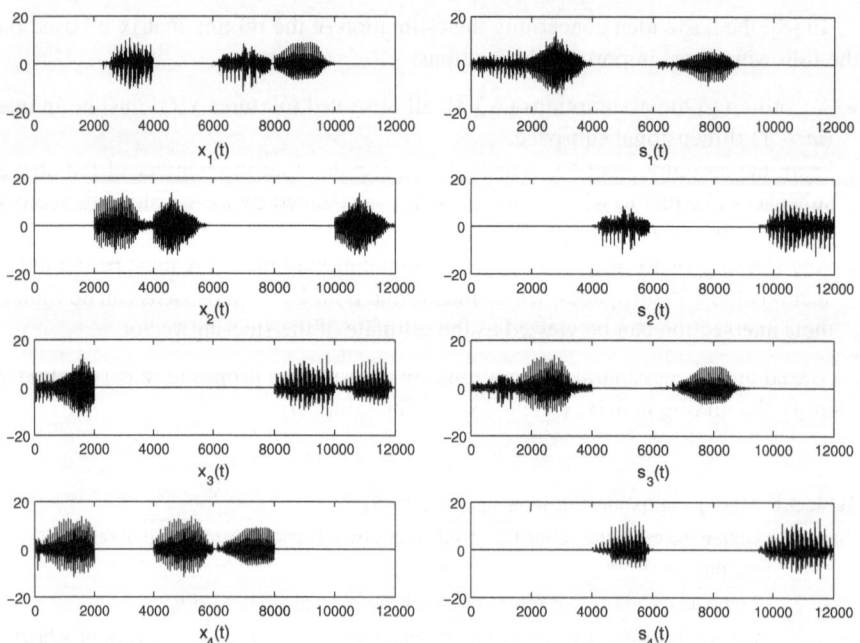

Fig. 3.1 Comparison between two sets of source signals. *Left column* source signal set $\{x_1(t), x_2(t), x_3(t), x_4(t)\}$ satisfying the assumption (A3.1). *Right column* source signal set $\{s_1(t), s_2(t), s_3(t), s_4(t)\}$ violating the assumption (A3.1)

and right columns in Fig. 3.1, respectively. It is easy to see that the source signals $\{x_1(t), x_2(t), x_3(t), x_4(t)\}$ satisfy all the above conditions (c1)–(c6) but the source signals $\{s_1(t), s_2(t), s_3(t), s_4(t)\}$ only meet the conditions (c2) and (c5). As a result, $\{s_1(t), s_2(t), s_3(t), s_4(t)\}$ cannot be separated from their mixtures by using the algorithm in [8]. From this example, we can see that the assumption (A3.1) significantly limits the application scope of this BSS algorithm and thus needs to be relaxed.

3.2.3 Mixing Matrix Estimation Under Local TF Sparsity

As shown in the above subsection, although the global sparsity assumption is acceptable for some kinds of signals, it is still too restrictive to satisfy in many practical applications. For these applications, the sparsity is satisfied only in some small or local time-frequency areas. Therefore, the local sparsity of source signals in time-frequency domain should be considered. To achieve this goal, the following restrictions are required:

(R3.1) The mixing matrix $\mathbf{A}$ is of full column rank.

(R3.2) Given $L = \lceil r/2 \rceil$, for any L source signals $x_{\theta 1}(t), \ldots, x_{\theta L}(t)$, there exist at least one sample interval $[P_0, P_0 + m]$ such that only these L sources $x_{\theta 1}(t), \ldots, x_{\theta L}(t)$ are active in this interval and any L columns in the matrix $\mathbf{X}(:, P_0 : P_0 + m)$ are linearly independent. Here, the matrix $\mathbf{X}(:, P_0 : P_0 + m)$ is defined as follows:

$$\mathbf{X}(:, P_0 : P_0 + m) = \begin{bmatrix} x_1(P_0) & x_1(P_0 + 1) & \cdots & x_1(P_0 + m) \\ x_2(P_0) & x_2(P_0 + 1) & \cdots & x_2(P_0 + m) \\ \vdots & \vdots & \vdots & \vdots \\ x_r(P_0) & x_r(P_0 + 1) & \cdots & x_r(P_0 + m) \end{bmatrix}_{r \times (m+1)}.$$

Clearly, the above restriction (R3.1) is easy to satisfy for any randomly generated non-underdetermined mixing matrix $\mathbf{A}$ whose column number is not less than the row number. Such a restriction has been exploited by numerous BSS algorithms. Besides, (R3.2) is a mild restriction about the sparsity of the source signals. It does not need the source signals to be sparse at all sample points but only a small number of sparse sample intervals are needed.

From (1.7), it holds that

$$\mathbf{Y}(:, P_0 : P_0 + m) = \mathbf{A} \cdot \mathbf{X}(:, P_0 : P_0 + m). \tag{3.4}$$

On the other hand, according to the restriction (R3.2), it is easy to see that if the number of active sources in the sample interval $[P_0, P_0 + m]$ is equal to L, then, rank$(\mathbf{X}(:, P_0 : P_0 + m))$ must be equal to L. Since $\mathbf{A}$ is of full column rank, it

Table 3.2 Mixing matrix identification algorithm based on local sparsity

Step 1	Find and collect all mixture matrices $\mathbf{Y}(:, P_0 : P_0 + m)$, whose rank is L
Step 2	Obtain C_r^L subspaces of dimension L after subspace clustering all mixed signal vectors $\mathbf{y}(t)$ in the mixture matrices $\mathbf{Y}(:, P_0 : P_0 + m)$
Step 3	If any C_{r-1}^{L-1} subspaces in these C_r^L subspaces have an intersection, then this intersection must be some steering vector $\mathbf{a}_i (i = 1, 2, \ldots, r)$ in the mixing matrix $\mathbf{A}$

follows from (3.4) that rank($\mathbf{Y}(:, P_0 : P_0 + m)$) is also L. Thus,

$$\text{span}\{\mathbf{Y}(:, P_0 : P_0 + m)\} = \text{span}\left\{\mathbf{a}_{\theta_1}, \ldots, \mathbf{a}_{\theta_L}\right\} \tag{3.5}$$

where $\mathbf{a}_{\theta_1}, \ldots, \mathbf{a}_{\theta_L}$ are the steering vectors associated with the active sources $x_{\theta_1}(t)$, $\ldots, x_{\theta_L}(t)$ in the sample interval $[P_0, P_0 + m]$. According to (3.5), one can obtain C_r^L subspaces via collecting $\mathbf{Y}(:, P_0 : P_0 + m)$ with rank L. Then, any steering vector $\mathbf{a}_i (i = 1, 2, \ldots, r)$ in the mixing matrix $\mathbf{A}$ must be the intersection among the C_{r-1}^{L-1} subspaces. Based on these viewpoints, an algorithm is derived to blindly identify the mixing matrix, which is shown in Table 3.2.

3.3 TFA-Based Methods for Source Recovery

After obtaining the mixing matrix, the remaining task is to recover the source signals from the observed mixtures. Clearly, if the mixing matrix $\mathbf{A}$ has full column rank, then the estimation of the source signals can be easily computed by left-multiplying the mixtures $\mathbf{y}(t)$ by the pseudo-inverse of the mixing matrix. However, if the mixing matrix is not of full column rank, e.g., in the underdetermined mixing system scenario, the recovery of source signals will be a challenging task. This section will discuss how to recover the source signals when the underdetermined mixing matrix is known.

3.3.1 Source Recovery Under Strong TF Sparsity

In early BSS works based on TFA [11, 13], the source signals are required to be TF-disjoint. That is, there is at most one active source at any TF point, which is a restrictive constraint. Although the TF-UBSS algorithms in [11, 13] can still work when there are some small overlapping among different sources in the TF domain, the separation performance will degrade at these overlapping points [18]. In order to overcome this drawback, some subspace based algorithms are developed in [18] for

TF-nondisjoint sources, where the quadratic TFDs are exploited. It is stated in [18] that the TFDs of different sources are allowed to overlap to some extent, given that the number of active sources at any TF point is strictly less than that of the mixtures and the column vectors of the mixing matrix are pairwise linearly independent.

In [18], the spatial diversity of source signals is combined with time-frequency analysis approaches to solve the problem of blind source recovery. For this goal, the STFD (spatial TFD) matrix of signals and the Cohen's class of TFD are utilized. From (1.7) and (3.3), it holds that

$$\mathbf{D_{yy}}(t, f) = \mathbf{A}\mathbf{D_{xx}}(t, f)\mathbf{A}^T. \tag{3.6}$$

The mixing matrix $\mathbf{A}$ is assumed to be underdetermined and known, and one needs to recover the source signals $\mathbf{x}(t)$. Moreover, the following assumptions are made:

(C3.1) The column vectors of $\mathbf{A}$ are pairwise linearly independent.

(C3.2) The number of active sources, denoted by L, at any TF point is strictly less than the number of mixtures, denoted by m.

Let us assume that at a given TF point (t', f'), the set of active sources are $\{x_{\alpha_1}, x_{\alpha_2}, \ldots, x_{\alpha_L}\}$. Denote $\tilde{\mathbf{x}}(t) = [x_{\alpha_1}(t), x_{\alpha_2}(t), \ldots, x_{\alpha_L}(t)]^T$ and $\tilde{\mathbf{A}} = [\mathbf{a}_{\alpha_1}, \mathbf{a}_{\alpha_2}, \ldots, \mathbf{a}_{\alpha_L}]$. Based on the assumption (C3.2), (3.6) can be reduced to the following form [18]

$$\mathbf{D_{yy}}(t', f') = \tilde{\mathbf{A}}\mathbf{D_{\tilde{x}\tilde{x}}}(t', f')\tilde{\mathbf{A}}^T. \tag{3.7}$$

Given that $\mathbf{D_{\tilde{x}\tilde{x}}}(t', f')$ is nonsingular, one can conclude from (3.7) that $\mathbf{D_{yy}}(t', f')$ and $\tilde{\mathbf{A}}$ have the same range. This is a very important observation which lays the foundation for the blind source recovery approaches proposed in [18].

Let $\mathbf{P} = \mathbf{I}_m - \mathbf{V}\mathbf{V}^H$, where $\mathbf{I}_m$ is the $m \times m$ identity matrix and $\mathbf{V}$ is the matrix formed by the L principal singular eigenvectors of $\mathbf{D_{yy}}(t', f')$. It is shown in [18] that

$$\begin{cases} \mathbf{P}\mathbf{a}_i = \mathbf{0}, & \text{for any } x_i \in \{x_{\alpha_1}, x_{\alpha_2}, \ldots, x_{\alpha_L}\} \\ \mathbf{P}\mathbf{a}_i \neq \mathbf{0}, & \text{for any } x_i \notin \{x_{\alpha_1}, x_{\alpha_2}, \ldots, x_{\alpha_L}\} \end{cases}. \tag{3.8}$$

Since the mixing matrix $\mathbf{A}$ is assumed to be known, (3.8) provides a criterion to determine which source signals are active at the time-frequency point (t, f). In other words, it can be used to identify the matrix $\tilde{\mathbf{A}}$ and the active sources at (t', f'). Clearly, by using the estimate of the mixing matrix $\tilde{\mathbf{A}}$, the TFD values of these active sources can be recovered using (3.7). Finally, the time-domain representations of these active sources can be obtained via a time-frequency synthesis procedure. Two TFA-based source recovery algorithms are proposed in [18].

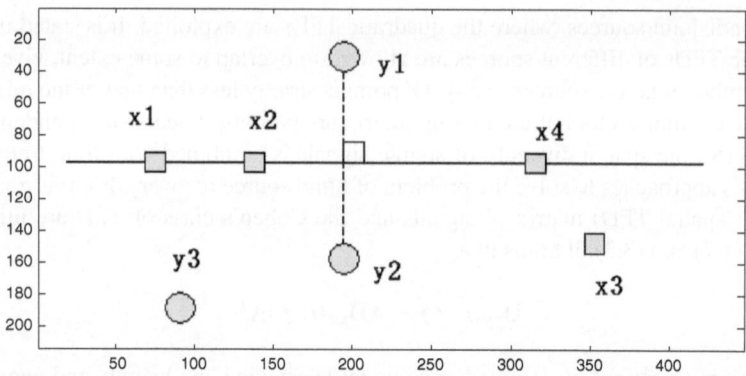

Fig. 3.2 An example of application where [18] fails

Discussion of the assumptions in [18]

The two equations in (3.8) play an essential role in the subspace-based algorithms in [18]. However, as shown in [7], the conditions (C3.1) and (C3.2) are not sufficient to guarantee (3.8). For instance, let us consider a mixing model, shown in Fig. 3.2, with four sources x_1, x_2, x_3, x_4 and three sensors y_1, y_2, y_3, where all four sources are omnidirectional. Assume that x_1, x_2, x_4 are placed along a line and the positions of y_1 and y_2 are symmetrical about this line. This topology results in $a_{11} = a_{21}$, $a_{12} = a_{22}$ and $a_{14} = a_{24}$, where a_{ij} is the (i, j)th entry of the mixing matrix $\mathbf{A}$. The source x_3 and sensor y_3 are placed in proper positions to ensure that the condition C3.1) holds (In fact, if x_3 and y_3 are randomly placed, the condition C3.1) holds with probability one). Because of the special structure of $\mathbf{A}$, the rank of the 3×3 matrix $[\mathbf{a}_1, \mathbf{a}_2, \mathbf{a}_4]$ is only two. Thus, there exist two nonzero constants λ_1 and λ_2 such that $\mathbf{a}_4 = \lambda_1 \mathbf{a}_1 + \lambda_2 \mathbf{a}_2$. Assume that at a given TF point, the set of active sources are $\{x_1, x_2\}$. From (3.8), we have $\mathbf{Pa}_1 = \mathbf{Pa}_2 = \mathbf{0}$. However, since $\mathbf{a}_4$ is a linear combination of $\mathbf{a}_1$ and $\mathbf{a}_2$, it yields $\mathbf{Pa}_4 = \mathbf{0}$, which contradicts with (3.8). This verifies the conclusion that the conditions (C3.1) and (C3.2) do not guarantee (3.8). In order to ensure (3.8), a stronger condition on the mixing matrix $\mathbf{A}$ must be satisfied, which is [7]

(A3.1') Any $m \times m$ submatrix of the mixing matrix $\mathbf{A}$ is of full rank.

On the other hand, it is important to point out that for the subspace-based algorithms in [18], the condition (C3.2) is essential. Otherwise, (3.8) does not satisfy. However, the condition (C3.2) is a restrictive constraint about the source signals in the time-frequency domain. In the next subsection, we will show a blind source recovery algorithm which relaxes the condition (C3.2) to that the active sources at any TF point can be as many as the mixed signals.

3.3.2 Source Recovery Under Relaxed TF Sparsity

We start from some necessary definitions and notations [7].

Definition 3.1 If a column vector is not a null vector, it is said to be *nonzero*.

Definition 3.2 Denote a $P \times Q$ matrix $\mathbf{Q}$ by $\mathbf{Q} = [\mathbf{q}_1, \mathbf{q}_2, \ldots, \mathbf{q}_P]^T$, where $\mathbf{q}_i^T$ $(i = 1, 2, \ldots, P)$ is the ith row vector of $\mathbf{Q}$. A matrix operator $\mathscr{P}$ for $\mathbf{Q}$ is defined as

$$\mathscr{P}(\mathbf{Q}) = \begin{bmatrix} \mathbf{q}_1^T \odot \mathbf{q}_2^T \\ \cdots \\ \mathbf{q}_1^T \odot \mathbf{q}_P^T \\ \mathbf{q}_2^T \odot \mathbf{q}_3^T \\ \cdots \\ \mathbf{q}_2^T \odot \mathbf{q}_P^T \\ \cdots \\ \mathbf{q}_i^T \odot \mathbf{q}_j^T \\ \cdots \\ \mathbf{q}_{P-1}^T \odot \mathbf{q}_P^T \end{bmatrix} \tag{3.9}$$

where $i < j$ and $\odot$ denotes the Hadamard product.

Definition 3.3 A matrix set $\mathbf{T}$ is composed of all $m \times m$ submatrices of $\mathbf{A}$ as follows:

$$\mathbf{T} = \left\{ \mathbf{T}^{(i)} \middle| \mathbf{T}^{(i)} = [\mathbf{a}_{\theta_1}, \mathbf{a}_{\theta_2}, \ldots, \mathbf{a}_{\theta_m}] \right\}. \tag{3.10}$$

Definition 3.4 A matrix set $\mathbf{H}$ is defined as

$$\mathbf{H} = \left\{ \mathbf{H}^{(k)} \middle| \mathbf{H}^{(k)} = \left(\mathbf{T}^{(i)} \right)^{-1} \mathbf{T}^{(j)}, \quad \mathbf{T}^{(i)}, \mathbf{T}^{(j)} \in \mathbf{T} \right\},$$

where $i \neq j$.

Given that the mixing matrix is known, the TFA-based blind source recovery algorithm to be introduced here depends on Assumption (A3.1') and the following two assumptions:

(A3.2') At any TF point, the number L of the active sources is less than or equal to the number m of the mixtures.

(A3.3') For any matrix $\mathbf{H}^{(k)} \in \mathbf{H}$, all of nonzero column vectors in $\mathscr{P}(\mathbf{H}^{(k)})$ are linearly independent.

Assumption (A3.1') is essential to the algorithm to be introduced, as well as the subspace-based algorithms in [18]. Assumption (A3.2') is less strict than Assumption

(C3.2) in [18] which requires the number of active sources at any TF point to be strictly less than that of mixtures. Assumption (A3.3′) is an additional condition. It is shown in [7] that if $m \geq 3$, Assumption (A3.3′) is satisfied with probability one for randomly generated mixing matrix. From Assumption (A3.1′) and the definitions of the matrix sets **T** and **H**, the following two lemmas can be derived [7].

Lemma 3.1 *Any element in the matrix sets* **T** *and* **H** *is of full column rank.*

Lemma 3.2 *For any matrix* $\mathbf{H}^{(k)}$ *from the set* **H**, *the number p of nonzero column vectors in* $\mathscr{P}\left(\mathbf{H}^{(k)}\right)$ *satisfies the relationship* $1 \leq p \leq \min\{m, r - m\}$, *where r stands for the number of sources and m denotes the number of mixtures.*

From the estimate of the mixing matrix **A**, one can compute the matrix set **T**. Also, the STFD matrices $\mathbf{D}_{yy}(t, f)$ of the measured mixed signals $y(t)$ can be computed from (3.3) at each TF point. In order to only preserve the TF points (t, f) with sufficient energy, a noise thresholding procedure proposed in [13] can be carried out: for each time-slice t_p of the TF domain, use the following criterion for all TF points (t_p, f_q) belonging to this time slice:

$$\text{If } \frac{\|\mathbf{D}_{yy}(t_p, f_q)\|}{\max_f \|\mathbf{D}_{yy}(t_p, f)\|} > \varepsilon_1, \text{ then keep}(t_p, f_q) \tag{3.11}$$

where ε_1 is a threshold (typically, $\varepsilon_1 = 0.05$). By applying the criterion (3.11), the TF points with negligible energy are removed.

To proceed, let us recall the definitions of auto-source and cross-source TF points [13]. At a TF point (t_a, f_a), if $D_{x_i x_i}(t_a, f_a)$ shows an energy concentration, then (t_a, f_a) is called an auto-source TF point of the source x_i. Similarly, at a TF point (t_c, f_c), if $D_{x_i x_j}(t_c, f_c)(i \neq j)$ shows an energy concentration, then (t_c, f_c) is called a cross-source TF point between two different sources x_i and x_j. It is known [13], [19] that the STFD matrix $\mathbf{D}_{xx}(t_a, f_a)$ of the sources is diagonal at an auto-source TF point (t_a, f_a). At a cross-source TF point (t_c, f_c), the matrix $\mathbf{D}_{xx}(t_c, f_c)$ is off-diagonal, i.e., its diagonal entries are zero.

It is shown in [13, 19] that trace $\left(\mathbf{W}\mathbf{D}_{yy}(t', f')\mathbf{W}^H\right) = $ trace $\left(\mathbf{D}_{xx}(t', f')\right)$, where **W** is the whitening matrix which is often estimated as an inverse square root of the covariance matrix of the measured mixtures, and the superscript ^{H} stands for complex conjugate transpose operation. Clearly, if (t', f') is a cross-source TF point, one can obtain trace $\left(\mathbf{W}\mathbf{D}_{yy}(t', f')\mathbf{W}^H\right) \approx 0$ by exploiting the off-diagonal structure of the source STFD matrix $\mathbf{D}_{xx}(t', f')$ at the cross-source TF point. Thus, the following criterion can be used to approximately identify the auto-source TF points and remove the cross-source TF points:

$$\text{If } \frac{\text{trace}\left(\mathbf{W}\mathbf{D}_{yy}(t', f')\mathbf{W}^H\right)}{\|\mathbf{W}\mathbf{D}_{yy}(t', f')\mathbf{W}^H\|} \geq \varepsilon_2, \text{ then } (t', f') \text{ is an auto-source TF point}$$

$$\tag{3.12}$$

where ε_2 is a threshold close to one (typically, $\varepsilon_2 = 0.8$). The performance of identifying the auto-source TF points depends on the value of $r - m$. That is, the smaller value $r - m$, the better performance on the identification of the auto-source TF points [13]. After this process, all auto-source TF points can be identified.

However, it is not straightforward to obtain the corresponding TFD values of each source signal as the active sources at different auto-source TF points may be different. To solve this problem, the following theorem is proposed [7].

Theorem 3.1 *For any auto-source TF point (t', f'), the following conclusions hold:*

(c3.1) If the assumptions (A3.1$'$) and (A3.2$'$) are satisfied, there exists a matrix $\mathbf{C} \in \mathbf{T}$ such that $\mathbf{C}^{-1}\mathbf{D}_{yy}(t', f')\mathbf{C}^{-T}$ is diagonal.

(c3.2) Under the assumptions (A3.1$'$)–(A3.3$'$), any matrix $\mathbf{C} \in \mathbf{T}$, which makes $\mathbf{C}^{-1}\mathbf{D}_{yy}(t', f')\mathbf{C}^{-T}$ become diagonal, contains all the steering vectors associated with the active sources at the TF point (t', f').

As discussed in [7], in the case of $m = 2$, the validity of the conclusion (c3.2) in Theorem 3.1 does not depend on Assumption (A3.3$'$).

From Theorem 3.1, one can obtain the TFD values of active sources at each auto-source TF point (t', f') through the following procedure. Firstly, in the set $\mathbf{T}$, the matrix $\mathbf{C}$, which makes $\mathbf{C}^{-1}\mathbf{D}_{yy}(t', f')\mathbf{C}^{-T}$ become diagonal, is found by

$$\mathbf{C} = \arg \max_{\mathbf{T}^{(k)} \in \mathbf{T}} \frac{\text{trace}\left[\left(\mathbf{T}^{(k)}\right)^{-1} \mathbf{D}_{yy}(t', f') \left(\mathbf{T}^{(k)}\right)^{-T}\right]}{\left\|\left(\mathbf{T}^{(k)}\right)^{-1} \mathbf{D}_{yy}(t', f') \left(\mathbf{T}^{(k)}\right)^{-T}\right\|}. \tag{3.13}$$

Secondly, from (3.7) and the conclusion (c3.2) in Theorem 1, the nonzero diagonal elements in the matrix $\mathbf{C}^{-1}\mathbf{D}_{yy}(t', f')\mathbf{C}^{-T}$ can be identified as the TFD values of the active sources at the TF point (t', f'). After the TFD values of a source signal at all auto-source TF points are obtained, the waveform of the source signal can be recovered via a TF synthesis procedure proposed in [39]. This UBSS algorithm is summarized in Table 3.3.

Simulation results

(1) Simulation 1

In this simulation, two examples are provided to illustrate the performance of the blind source recovery algorithms in [7, 18]. To simplify the illustration and

Table 3.3 Source recovery using the UBSS algorithm in [7]

Step 1	Estimate the mixing matrix $\mathbf{A}$ using a method in [18, 22] or [23]
Step 2	Compute the matrix set $\mathbf{T}$ and the STFD matrices $\mathbf{D}_{yy}(t, f)$ of the mixtures at all TF points by (3.3)
Step 3	Remove the TF points with negligible energy by (3.11)
Step 4	Identify the auto-source TF points by (3.12)
Step 5	Obtain the TFD values of the active sources at the auto-source TF points by (3.13)
Step 6	Estimate the source signals using the TF synthesis algorithm in [39]

comparison of performance of these algorithms, the linear frequency modulation (LFM) signals whose instantaneous frequencies vary linearly with time are used as source signals in the simulations. The source signals are mixed by an instantaneous mixing matrix of dimension $m \times r$.

In the first example, $m = 3$ sensors are used to receive $r = 4$ LFM source signals $x_1(t)$, $x_2(t)$, $x_3(t)$ and $x_4(t)$, each of which has 512 samples. The normalized frequency ranges of $x_1(t)$, $x_2(t)$, $x_3(t)$ and $x_4(t)$ are $(0, 0.1)$, $(0.5, 0)$, $(0.3, 0.5)$ and $(0.25, 0.23)$, respectively. Clearly, the TFDs of the original sources are not disjoint and there are $L = 2$ active sources at some TF points. The mixing matrix $\mathbf{A}$ is

$$\mathbf{A} = \begin{bmatrix} 0.8326 & 0.6255 & 0.5975 & 0.5186 \\ 0.5167 & 0.7800 & 0.3328 & 0.7317 \\ 0.1994 & 0.0183 & 0.7295 & 0.4423 \end{bmatrix}$$

which is randomly generated. The TFDs of the reconstructed sources are shown in Fig. 3.3, where the sub-figures in the left column are the TFDs of sources recovered by the subspace-based quadratic TF-UBSS algorithm in [18] and those in the right column are the TFDs of sources recovered by the algorithm in [7]. We can see that both algorithms achieve good performance in source recovery.

In the second example, based on the simulation parameters used in Example 1, we assume that the third sensor is removed. This results in the following mixing matrix:

$$\mathbf{A} = \begin{bmatrix} 0.8326 & 0.6255 & 0.5975 & 0.5186 \\ 0.5167 & 0.7800 & 0.3328 & 0.7317 \end{bmatrix}$$

i.e., the mixing system has four sources and two sensors. For this case, there are $L = 2$ active sources at some TF points. Since the subspace-based methods in [18] require that the number of active sources at any TF point should be strictly less than that of sensors, they do not work in this example. In contrast, the algorithm in [7] still works well in this case and successfully separate the four source signals from their mixtures. The waveforms and TFDs of the recovered sources are shown in Fig. 3.4.

(2) Simulation 2

In the second simulation, we consider a mixing system with three sources and two sensors. The three source signals are speech signals and the 2×3 mixing matrix is randomly generated from the distribution $\mathcal{N}(0, 1)$. The algorithm in [7] is used to separate the three source signals from their mixtures measured by the two sensors. In Fig. 3.5, the top three subplots (a)–(c) show the time–domain representations of the original source signals, the middle two subplots (d)–(e) represent the waveforms of the two mixed signals and the bottom three subplots (f)–(h) represent the recovered source signals.

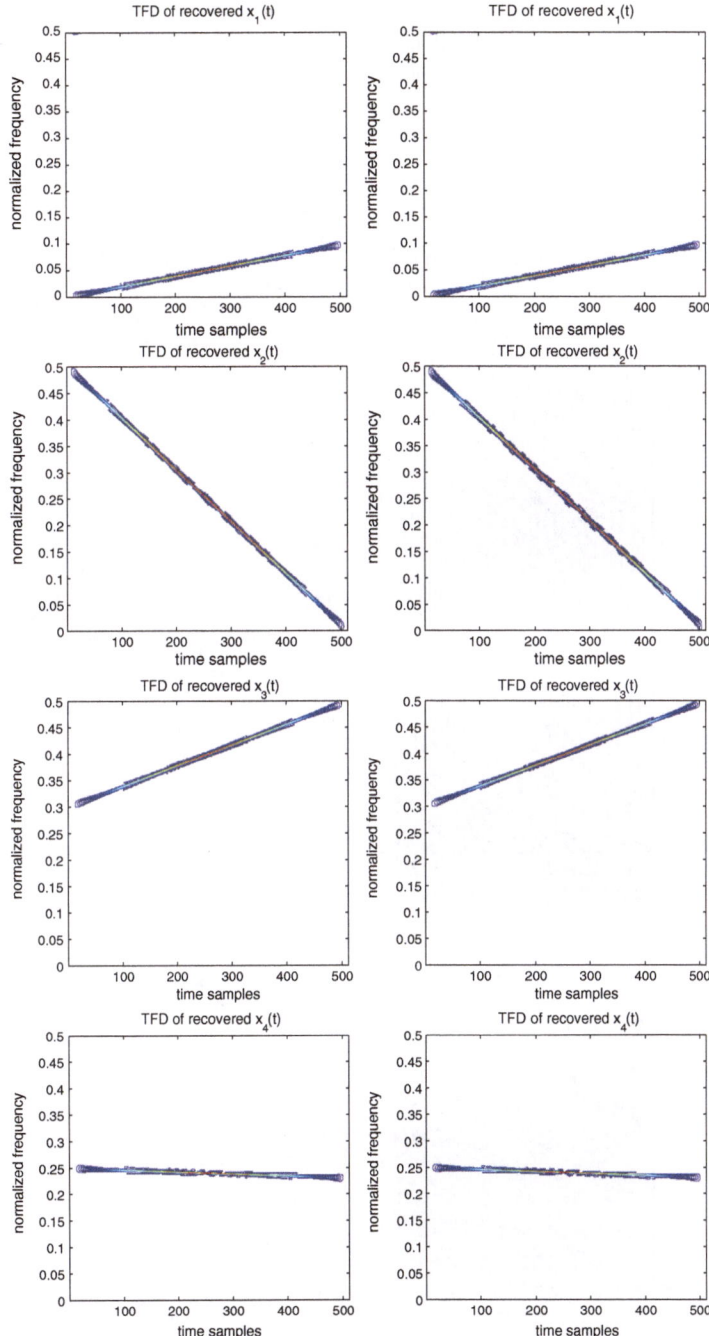

Fig. 3.3 TF signatures of recovered source signals, where $m = 3$ and $r = 4$. *Left* TF signatures of the recovered source signals by the quadratic TF-UBSS algorithm in [18]. *Right* TF signatures of the recovered source signals by the algorithm in [7]

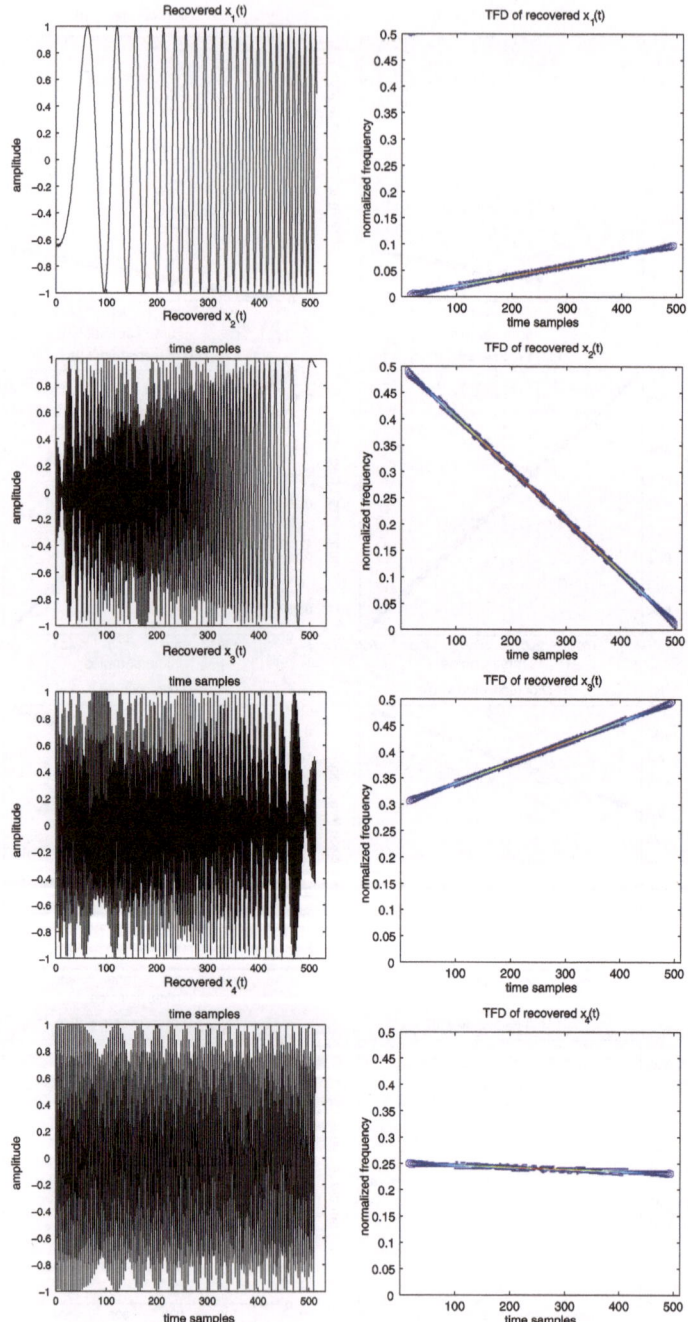

Fig. 3.4 Recovered source signals by the algorithm in [7], where the mixing system has 2 sensors and 4 sources. *Left* Waveforms of recovered signals. *Right* TF signatures of recovered signals

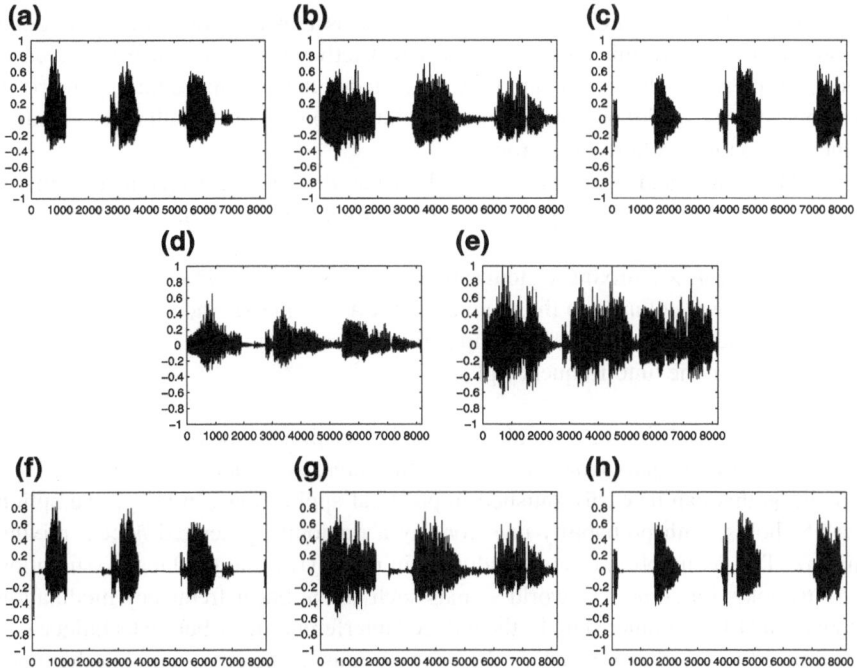

Fig. 3.5 **a–c** Original source signals; **d–e** Mixed signals; **f–h** Recovered source signals by the algorithm in [7]

3.3.3 Recovery of Non-sparse Dependent Sources

Suppose that the underdetermined mixing system is known (or can be estimated by other methods). In order to recover the source signals from their mixtures, most existing approaches rely on the assumption that the sources are sparse in time domain, or other transformation domains such as the TF domain. However, many real-world signals such as majority of communication signals do not possess the sparsity property. Hence, it is important to develop blind source recovery algorithms that do not impose any sparsity constraint on the sources.

A small number of approaches have been proposed for the blind separation of non-sparse sources [2–5, 16, 26, 27]. In [26, 27], the UBSS problem is tackled under the assumption that some of the sources are nonstationary signals and the others are stationary signals. It is shown that the difference between the nonstationary sources and stationary sources can be exploited to separate the former from the mixtures. The stationary sources are treated as noise signals and thus cannot be recovered. In [2–5], a sequential blind extraction technique is utilized to estimate the source signals mixed by a mixing matrix **A** having some special properties. More specifically, in order to estimate a single source signal, it is required that there should exist one column in the mixing matrix $\mathbf{A}$ which cannot be represented as a linear combination of all other columns. Obviously, this condition is very restrictive in the underdetermined case. For example, any column of a randomly generated underdetermined mixing matrix $\mathbf{A}$ can be expressed, with probability one, as a linear combination of all other columns since $\mathbf{A}$ has more columns than rows.

In [16], Peng and Xiang propose a TFA-based source recovery algorithm for non-sparse sources, which is built upon the following hypotheses:

(A3.4) The $m \times r$ mixing system satisfies $m \geq 3$ and $r \leq 2m - 1$.

(A3.5) Any m columns in the mixing matrix $\mathbf{A}$ are linearly independent.

(A3.6) There is *almost* no superimposition between auto-source point and cross-source point in the time-frequency plane.

Assumption (A3.4) implies that the proposed algorithm can be used to extract up to $2m - 1$ source signals from $m(m \geq 3)$ mixtures. Assumption (A3.5) is a mild condition that can be easily satisfied in practical applications. In fact, Assumption (A3.5) holds, with probability one, for any a randomly generated $m \times r$ mixing matrix. This assumption has been adopted in many UBSS algorithms. Assumption (A3.6) holds for some real-world signals such as the linear frequency modulation signals in [13]. For other signals, the reduced interference distribution technique can be utilized to significantly reduce the contributions of the cross-terms in the time-frequency plane, i.e., to make the off-diagonal elements in $\mathbf{D}_{\mathbf{xx}}(t, f)$ negligible at any auto-source point (t, f) [19]. This ensures Assumption (A3.6) to be satisfied. It should also be pointed out that even though a small number of source TFD values are erroneously estimated due to small overlapping between auto-source points and cross-source points, the performance deterioration of the TF-based UBSS algorithm will be very limited, as illustrated in [13].

To introduce this algorithm, some important matrices which are related to the mixing matrix $\mathbf{A}$ are defined as follows.

Definition 3.5 An $m \times m$ submatrix $\bar{\mathbf{A}}$ of the mixing matrix $\mathbf{A}$ is defined as

$$\bar{\mathbf{A}} \triangleq [\mathbf{a}_1, \mathbf{a}_2, \ldots, \mathbf{a}_m] \tag{3.14}$$

where $\mathbf{a}_i$ is the ith column in $\mathbf{A}$, $i = 1, 2, \ldots, m$.

According to Assumption (A3.5), it is obvious that $\bar{\mathbf{A}}$ is invertible.

Definition 3.6 An $m \times r$ matrix $\mathbf{U}$ is defined as

$$\mathbf{U} \triangleq \bar{\mathbf{A}}^{-1}\mathbf{A}. \tag{3.15}$$

Definition 3.7 Define

Table 3.4 Source recovery algorithm in [16]

Step 1	Use a blind identification algorithm in [22, 23] to estimate the mixing matrix $\mathbf{A}$
Step 2	Construct the matrices $\bar{\mathbf{A}}$ by (3.14), $\mathbf{U}$ by (3.15) and $\bar{\mathbf{U}}$ by (3.16)
Step 3	Compute the STFD matrix $\mathbf{D}_{\mathbf{yy}}(t, f)$ of the mixed signals $\mathbf{y}(t)$. Then, obtain the matrix $\mathbf{B}(t, f)$ by (3.17)
Step 4	For a given time slice t_p, remove the TF points satisfying $$\frac{\|\mathbf{D}_{\mathbf{yy}}(t_p, f_q)\|}{\max_f \|\mathbf{D}_{\mathbf{yy}}(t_p, f)\|} \leq \varepsilon_1$$ where ε_1 is a threshold (typically, $\varepsilon_1 = 0.05$)
Step 5	The TF points (t', f') satisfying $$\frac{\|\bar{\mathbf{U}}\bar{\mathbf{U}}^{\#}\mathrm{vec}(\mathbf{B}^T(t',f'))-\mathrm{vec}(\mathbf{B}^T(t',f'))\|}{\|\mathrm{vec}(\mathbf{B}^T(t',f'))\|} \leq \varepsilon_2$$ are selected as auto-source TF points. Here, the superscript $^{\#}$ denotes the Moore-Penrose's pseudo-inverse operation, $\mathrm{vec}(\cdot)$ stands for the vectorization operation on a matrix, and ε_2 is a threshold (typically, $\varepsilon_2 = 0.1$)
Step 6	Estimate the TFD values of the sources at the auto-source TF point (t', f') by $$\left[D_{x_1 x_1}(t', f'), \ldots, D_{x_r x_r}(t', f')\right]^T = \bar{\mathbf{U}}^{\#}\mathrm{vec}\left(\mathbf{B}^T(t', f')\right)$$
Step 7	After collecting the TFD values of active sources at all auto-source TF points, one can recover the original waveforms of source signals using the TF synthesis procedures in [39]

$$\bar{\mathbf{U}} \triangleq \begin{bmatrix} \mathbf{e}_1 & \mathbf{0} & \cdots & \mathbf{0} & u_{1(m+1)}\mathbf{u}_{m+1} & \cdots & u_{1r}\mathbf{u}_r \\ \mathbf{0} & \mathbf{e}_2 & \cdots & \mathbf{0} & u_{2(m+1)}\mathbf{u}_{m+1} & \cdots & u_{2r}\mathbf{u}_r \\ \vdots & \vdots & \ddots & \vdots & \vdots & \vdots & \vdots \\ \mathbf{0} & \mathbf{0} & \cdots & \mathbf{e}_m & u_{m(m+1)}\mathbf{u}_{m+1} & \cdots & u_{mr}\mathbf{u}_r \end{bmatrix} \quad (3.16)$$

where $\mathbf{0}$ is the m-dimensional zero vector, $\mathbf{e}_i$ is an m-dimensional column vector in which the ith element is equal to 1 and all other elements are zero, $u_{ij}(i = 1, \ldots, m, j = m+1, \ldots, r)$ is the (i, j)th entry of $\mathbf{U}$, and $\mathbf{u}_j$ is the jth column of $\mathbf{U}$.

Definition 3.8 Given any time-frequency point (t, f), define

$$\mathbf{B}(t, f) \triangleq \bar{\mathbf{A}}^{-1}\mathbf{D}_{\mathbf{yy}}(t, f)\bar{\mathbf{A}}^{-T}. \quad (3.17)$$

Based on the notations given in the above definitions, an effective TFA-based blind source recovery algorithm is developed in [16]. It consists of seven steps and is summarized in Table 3.4.

Simulation result

A simulation example is provided to illustrate the effectiveness of the algorithm in [16]. In the simulation, we use $m = 3$ sensors to receive $r = 4$ LFM source signals $x_1(t)$, $x_2(t)$, $x_3(t)$ and $x_4(t)$ in the absence of noise, where each measured signal has 512 samples. The normalized frequency ranges of $x_1(t)$, $x_2(t)$, $x_3(t)$ and $x_4(t)$ are $[0, 0.5]$, $[0.24, 0.26]$, $[0.5, 0]$, $[0.1, 0.4]$, respectively. Figure 3.6 shows the

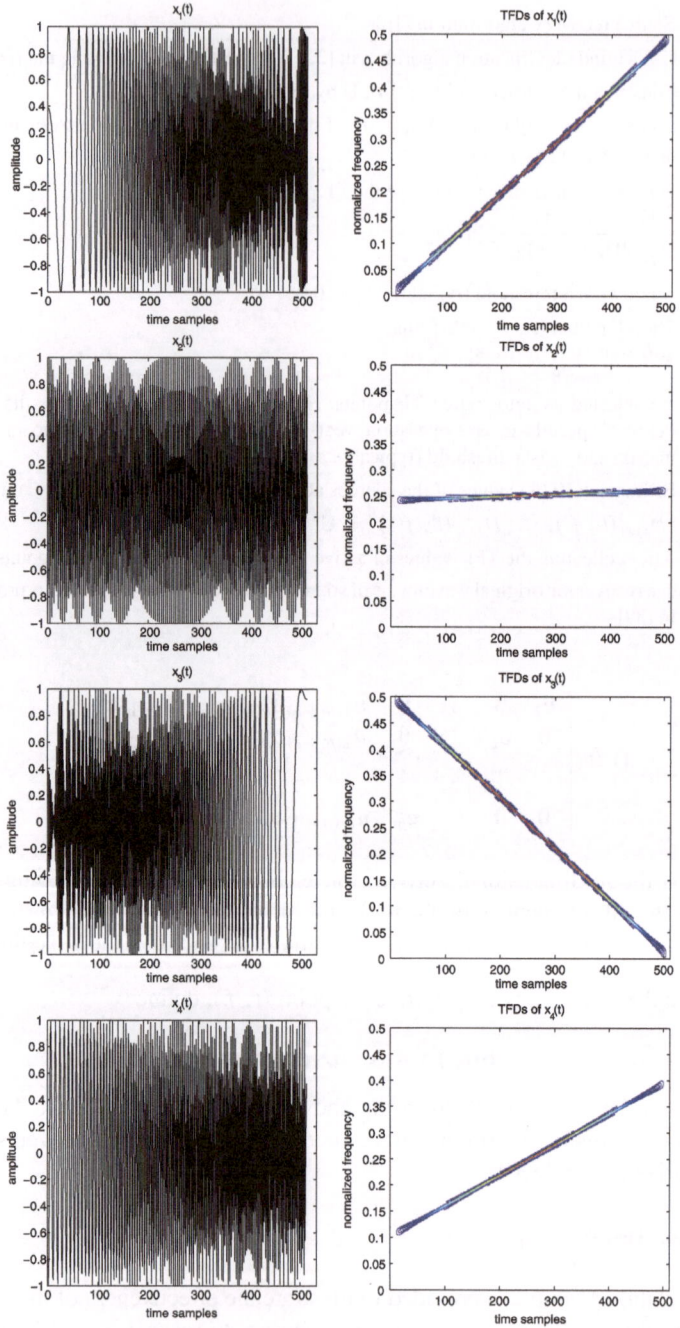

Fig. 3.6 The waveforms and TF distributions of the original source signals. *Left* waveforms; *Right* TF distributions

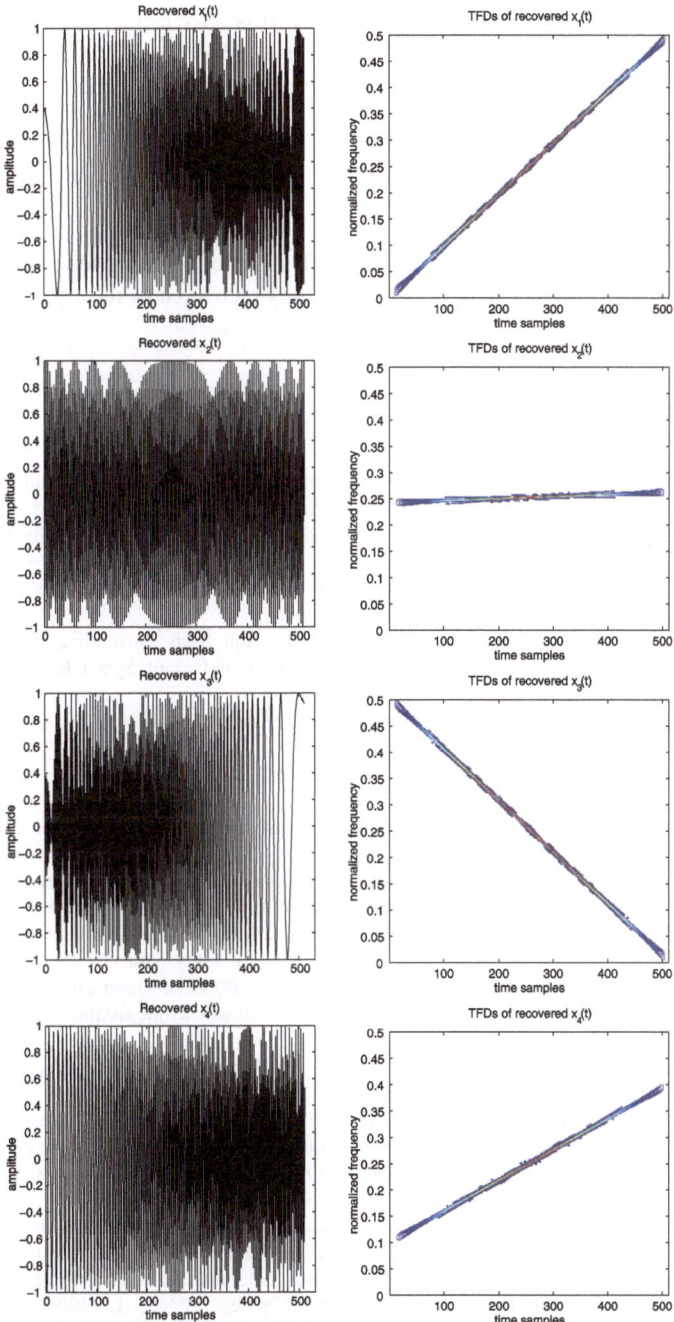

Fig. 3.7 The waveforms and TF distributions of the source signals reconstructed by the algorithm in [16]. *Left* waveforms; *Right* TF distributions

waveforms and TFDs of these source signals. It is easy to see that all sources are active in the neighbourhood of the TF point ($t = 256$, $f = 0.25$). The four source signals are mixed by the following randomly generated mixing matrix:

$$\mathbf{A} = \begin{bmatrix} 0.8326 & 0.6255 & 0.5975 & 0.5186 \\ 0.5167 & 0.7800 & 0.3328 & 0.7317 \\ 0.1994 & 0.0183 & 0.7295 & 0.4423 \end{bmatrix}.$$

Figure 3.7 shows the waveforms and the TFDs of the reconstructed source signals. Clearly, all source signals have been successfully recovered by the algorithm in [16].

References

1. R. Gribonval, S. Lesage, A survey of sparse component analysis for blind source separation: principles, perspectives, and new challenges. Proc. ESANN **1**, 323–330 (2006)
2. Y. Li, J. Wang, J.M. Zurada, Blind extraction of singularly mixed source signals. IEEE Trans. Neural Netw. **11**(6), 1413–1422 (2000)
3. Y. Li, J. Wang, Sequential blind extraction of instantaneously mixed sources. IEEE Trans. Signal Process. **50**(5), 997–1006 (2002)
4. Y. Li, J. Wang, A. Cichocki, Blind source extraction from convolutive mixtures in ill-conditioned multi-input multi-output channels. IEEE Trans. Circuits Syst. I: Regulat Pap. **51**(9), 1814–1822 (2004)
5. Y. Li, J. Wang, A network model for blind source extraction in various ill-conditioned cases. Neural Netw. **18**(10), 1348–1356 (2005)
6. P. Bofill, M. Zibulevsky, Underdetermined blind source separation using sparse representations. Signal Process. **81**(11), 2353–2362 (2001)
7. D. Peng, Y. Xiang, Underdetermined blind source separation based on relaxed sparsity condition of sources. IEEE Trans. Signal Process. **57**(2), 809–814 (2009)
8. P. Georgiev, F. Theis, A. Cichocki, Sparse component analysis and blind source separation of underdetermined mixtures. IEEE Trans. Neural Netw. **16**(4), 992–996 (2005)
9. N. Linh-Trung, A. Aissa-El-Bey, K. Abed-Meraim, A. Belouchrani, Underdetermined blind source separation of non-disjoint nonstationary sources in time-frequency domain, in *Proceedings of the 8th International Symposium Signal Processing Applications*, pp. 46–49 (2005)
10. O. Yilmaz, S. Rickard, Blind separation of speech mixtures via time-frequency masking. IEEE Trans. Signal Process. **52**(7), 1830–1847 (2004)
11. A. Jourjine, S. Rickard, O. Yilmaz, Blind separation of disjoint orthogonal signals: demixing N sources from 2 mixtures, in *Proceedings of the 2000 IEEE International Conference Acoustics, Speech, and Signal Processing*, vol. 5, pp. 2985–2988 (2000)
12. L.T. Nguyen, A. Belouchrain, K. Abed-Meraim, B. Boashash, Separating more sources than sensors using time-frequency distributions, in *Proceedings of the 6th International Symposium Signal Processing and Its Application*, pp. 583–586 (2001)
13. N. Linh-Trung, A. Belouchrani, K. Abed-Meraim, Separating more sources than sensors using time-frequency distributions. EURASIP J. Appl. Signal Process. **17**, 2828–2847 (2005)
14. Y. Zhang, M.G. Amin, Blind separation of nonstationary sources based on spatial time-frequency distributions. EURASIP J. Appl. Signal Process. **2006**, 1–13 (2006)
15. D. Peng, Y. Xiang, H. Trinh, A new blind method for separating M+1 sources from M mixtures. Comput. Math. Appl. **60**(7), 1829–1839 (2010)
16. D. Peng, Y. Xiang, Underdetermined blind separation of non-sparse sources using spatial time-frequency distributions. Digit. Signal Process. **20**(2), 581–596 (2010)

17. L. Cohen, *Time-Frequency Analysis* (Prentice-Hall, NJ, 1995)
18. A. Aissa-El-Bey, N. Linh-Trung, K. Abed-Meraim, A. Belouchrani, Y. Grenier, Underdetermind blind separation of nondisjoint sources in the time-frequency domain. IEEE Trans. Signal Process. **55**(3), 897–907 (2007)
19. A. Belouchrani, M.G. Amin, Blind source separation based on time-frequency signal representations. IEEE Trans. Signal Process. **46**(11), 2888–2897 (1998)
20. C. Wei, W.L. Woo, S.S. Dlay, Nonlinear underdetermined blind signal separation using Bayesian neural network approach. Digit. Signal Process. **17**(1), 50–68 (2007)
21. V. Zarzoso, Second-order criterion for blind source extraction. Electron. Lett. **44**(22), 1327–1328 (2008)
22. A. Ferreol, L. Albera, P. Chevalier, Fourth-order blind identification of underdetermined mixtures of sources (FOBIUM). IEEE Trans. Signal Process. **53**(5), 1640–1653 (2005)
23. L.D. Lathauwer, J. Castaing, J.F. Cardoso, Fourth-order cumulant-based blind identification of underdetermined mixtures. IEEE Trans. Signal Process. **55**(6), 2965–2973 (2007)
24. A.T. Cemgil, C. Fevotte, S.J. Godsill, Variational and stochastic inference for Bayesian source separation. Digit. Signal Process. **17**(5), 891–913 (2007)
25. Q. Lin, F. Yin, T. Mei, H. Liang, A blind source separation based method for speech encryption. IEEE Trans. Circuits Syst. I Regul. Pap. **53**(6), 1320–1328 (2006)
26. Y. Deville, M. Benali, F. Abrard, Differential source separation for underdetermined instantaneous or convolutive mixtures: concept and algorithms. Signal Process. **84**(10), 1759–1776 (2004)
27. J. Thomas, Y. Deville, S. Hosseini, Differential fast fixed-point algorithms for underdetermined instantaneous and convolutive partial blind source separation. IEEE Trans. Signal Process. **55**(7), 3717–3729 (2007)
28. D. Luengo, I. Santamaria, J. Ibanez, L. Vielva, C. Pantaleon, A fast blind SIMO channel identification algorithm for sparse sources. IEEE Signal Process. Lett. **10**(5), 148–151 (2003)
29. S. Xie, L. Yang, J.-M. Yang, G. Zhou, Y. Xiang, A time-frequency approach to underdetermined blind source separation. Neural Netw. Learn. Syst. **23**(2), 306–316 (2012)
30. P.S. Bradley, O.L. Mangasarian, k-plane clustering. J. Global Optim. **16**(1), 23–32 (2000)
31. R. Vidal, Subspace clustering. IEEE Signal Process. Mag. **28**(2), 52–68 (2011)
32. F. Abrard, Y. Deville, P. White, From blind source separation to blind source cancellation in the underdetermined case: a new approach based on time-frequency analysis, in *Proceedings of the 3rd International Conference Independent Component Analysis Signal Separation*, pp. 734–739 (2001)
33. F. Abrard, Y. Deville, Blind separation of dependent sources using the time-frequency ratio of mixtures approach, in *Proceedings of the 7th International Symposium on Signal Processing Applications*, pp. 1–4 (2003)
34. F. Abrard, Y. Deville, A time-frequency blind signal separation method applicable to underdetermined mixtures of dependent sources. Signal Process. **85**(7), 1389–1403 (2005)
35. A. Javanmard, P. Pad, M. Babaie-Zadeh, C. Jutten, Estimating the mixing matrix in underdetermined sparse component analysis (SCA) using consecutive independent component analysis (ICA), in *Proceedings of the 16th European Signal Processing Conference*, 2008
36. D.L. Donoho, M. Elad, Optimally sparse representation in general (non-orthogonal) dictionaries via ℓ1 minimization. Proc. Natl. Acad. Sci. **100**, 2197–2202 (2003)
37. M.S. Karoui, Y. Deville, S. Hosseini, A. Ouamri, Blind spatial unmixing of multispectral images: new methods combining sparse component analysis, clustering and non-negativity constraints. Pattern Recogn. **45**, 4263–4278 (2012)
38. J.F. Cardoso, A. Souloumiac, Blind beamforming for non-Gaussian signals. IEE Proc. F **140**(6), 362–370 (1993)
39. G.F. Boudreaux-Bartels, T.W. Parks, Time-varying filtering and signal estimation using Wigner distribution synthesis techniques. IEEE Trans. Acoust. Speech Signal Process. **ASSP–34**(3), 442–451 (1986)
40. L. Sorber, M.V. Barel, L.D. Lathauwer, Tensorlab v1.0, Available online, Feb. 2013. http://esat.kuleuven.be/sista/tensorlab/

Chapter 4
Dependent Component Analysis Using Precoding

Abstract Most existing methods for dependent component analysis only work under the condition that the source signals are nonnegative and/or sparse [1, 2]. Unfortunately, the signals in many real-world applications such as wireless communication systems are neither nonnegative nor sparse. In this chapter, three precoding based methods are presented to separate nonnegative and non-sparse but spatially correlated sources. The precoding based methods take advantage of the fact that in some applications, the source signals are accessible before being mixed up. For example, in a wireless communication system, the user signals at the transmission end are accessible prior to being transmitted to the receiver. This provides an opportunity to preprocess them before transmission such that BSS can be achieved at the receiver. Different from the method in [3], the precoding based methods do not impose any condition on the time-frequencys distributions of the sources.

Keywords Precoding · Mutually correlated sources · Dependent component analysis · Singular value · Singular vector

4.1 Concept of Precoding Based Dependent Component Analysis

Let $\mathbf{s}(t) = [s_1(t), s_2(t), \ldots, s_r(t)]^T$ be r nonnegative and non-sparse but mutually correlated signals, where the superscript ^{T} denotes transpose. If they are mixed up by an $m \times r, m \geq r$ instantaneous channel system $\mathbf{A}$, the current technologies cannot retrieve them from their mixtures. To tackle this problem, it is proposed to preprocess these signals before they are transmitted out. Specifically, before transmission, they are passed through a group of precoders $p_1(z), p_2(z), \ldots, p_r(z)$, respectively. Denote the coded signals by $\mathbf{x}(t) = [x_1(t), x_2(t), \ldots, x_r(t)]^T$ and define the ith precoder by

$$p_i(z) = \sum_{l=0}^{L} p_{i,l} z^{-l} \qquad (4.1)$$

Y. Xiang et al., *Blind Source Separation*,
SpringerBriefs in Signal Processing, DOI 10.1007/978-981-287-227-2_4

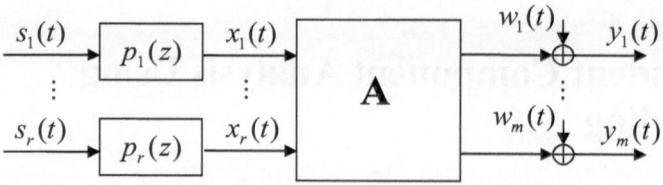

Fig. 4.1 Block diagram of the system model using precoders

where L is the highest order of the precoders. Then the coded signals can be written as

$$x_i(t) = p_i(z)s_i(t)$$
$$= p_{i,0}s_i(t) + p_{i,1}s_i(t-1) + \cdots + p_{i,L}s_i(t-L) \qquad (4.2)$$

where $i = 1, 2, \ldots, r$. The purpose of applying precoders to the source signals is that with properly designed precoders, the coded signals $\mathbf{x}(t)$ have certain spatial diversity.

Then, the coded signals $\mathbf{x}(t)$ are transmitted over the instantaneous channel system $\mathbf{A}$. At the receiving end, the received signals are measured in the presence of additive noise signals $\mathbf{w}(t) = [w_1(t), w_2(t), \ldots, w_m(t)]^T$. We denote the measured signals by $\mathbf{y}(t) = [y_1(t), y_2(t), \ldots, y_m(t)]^T$. Figure 4.1 shows the block diagram of the system model. Mathematically, we have $\mathbf{y}(t) = \mathbf{A}\mathbf{x}(t) + \mathbf{w}(t)$. Taking advantage of the spatial diversity of $\mathbf{x}(t)$ introduced by the precoders, one can retrieve the coded signals $\mathbf{x}(t)$ from $\mathbf{y}(t)$ without knowing the mixing matrix $\mathbf{A}$.

The system model considered here looks similar to the one in (1.2) with the difference being that $\mathbf{x}(t)$ here are the coded signals rather than the original source signals. After estimating $\mathbf{x}(t)$, further action is needed to recover the source signals $\mathbf{s}(t)$ from $\mathbf{x}(t)$ with the help of information about the precoders.

4.2 Precoding Based Time-Domain Method

Consider the system in Fig. 4.1 in the absence of noise, i.e., $w_i(t) = 0$, $i = 1, 2, \ldots, m$. In this case, the system model reduces to $\mathbf{y}(t) = \mathbf{A}\mathbf{x}(t)$. Let $\mathbf{R}_{ss}(\tau)$ be the autocorrelation matrix of $\mathbf{s}(t)$ at time lag τ, i.e.,

$$\mathbf{R}_{ss}(\tau) = E\left(\mathbf{s}(t)\mathbf{s}(t-\tau)^T\right) \qquad (4.3)$$

where the (i, j)th entry of $\mathbf{R}_{ss}(\tau)$ is $r_{ij}(\tau) = E\left(s_i(t)s_j(t-\tau)\right)$. Assume

(A4.1) The number of mixtures is not less than that of the sources, i.e., $m \geq r \geq 2$, and the channel matrix $\mathbf{A}$ is full column rank.

(A4.2) The source signals $s_i(t)$, $i = 1, 2, \ldots, r$ are stationary and mutually correlated, and

$$\mathbf{R_{ss}}(\tau) = \mathbf{0} \quad \text{for } |\tau| \geq M \tag{4.4}$$

where M is some positive integer and $\mathbf{R_{ss}}(0)$ is nonsingular.

(A4.3) All precoder coefficients $p_{i,l}$, $i = 1, 2, \ldots, r$ and $l = 1, 2, \ldots, L$ are known at receiver.

We would like to note here that (4.4) implies that the source signals can be temporally colored but the length of color, $M - 1$, is finite. The upper bound of M could be known in some applications. Otherwise, one needs to estimate it. Since each transmitter does not have information of the source signals from other transmitters, M cannot be estimated directly from $\mathbf{R_{ss}}(\tau)$ at the transmission side. Instead, it can be estimated at the receiver through an additional procedure of transmitting the source sequences without using precoders. Since $x_i(t) = s_i(t)$ in this case, the correlation matrix of the mixed signals at time lag τ is $\mathbf{R_{yy}}(\tau) = \mathbf{A}\mathbf{R_{ss}}(\tau)\mathbf{A}^T$, which leads to $\mathbf{R_{ss}}(\tau) = \mathbf{A}^\#\mathbf{R_{yy}}(\tau)(\mathbf{A}^\#)^T$, where the superscript $^\#$ stands for pseudo-inverse operation. Since $\mathbf{R_{ss}}(\tau) = \mathbf{0}$ if and only if $\mathbf{R_{yy}}(\tau) = \mathbf{0}$, M can be estimated from $\mathbf{R_{yy}}(\tau)$.

Precoder Design

Given $K \geq M$, it holds from (4.4) that

$$\mathbf{R_{ss}}(K) = \mathbf{0}. \tag{4.5}$$

Choose the highest order of the precoders to be

$$L = (4r - 1)K \geq (4r - 1)M$$

and design the precoder coefficients as follows [4]:

$$p_{i,l} \begin{cases} \neq 0 & \text{if } l = (2r - 2i)K, \ (2r + 2i - 1)K \\ = 0 & \text{for other } l \end{cases} \tag{4.6}$$

where $i = 1, 2, \ldots, r$. It can be seen from (4.6) that each precoder has two nonzero coefficients. One can simply select the nonzero coefficients of the precoders as follows:

$$p_{i,l_i} = 1 \quad \text{and} \quad p_{i,l_i'} = \rho_i \tag{4.7}$$

for all $1 \leq i \leq r$, where $\rho_1, \rho_2, \ldots, \rho_r$ are nonzero real numbers and

$$l_i = (2r - 2i)K \tag{4.8}$$

$$l_i' = (2r + 2i - 1)K. \tag{4.9}$$

From (4.2), (4.6) and (4.7), it follows

$$x_i(t) = s_i(t - l_i) + \rho_i s_i(t - l_i')$$ (4.10)

where $i = 1, 2, \ldots, r$.

Let

$$\tau_q = (4q - 1)K$$ (4.11)

where $q = 1, 2, \ldots, r$. Then, based on the above precoders, one has the following two lemmas [4]:

Lemma 4.1 *Suppose* $i, q \in \{1, 2, \ldots, r\}$. *For a given time lag* τ_q, *the following equation holds:*

$$E\left(x_i(t)x_i(t - \tau_q)\right) = \begin{cases} \rho_q r_{qq}(0) & \text{if} \quad i = q \\ 0 & \text{if} \quad i \neq q. \end{cases}$$ (4.12)

Lemma 4.2 *Suppose* $i, j, q \in \{1, 2, \ldots, r\}$ *and* $i \neq j$. *For a given time lag* τ_q, *it holds that*

$$E\left(x_i(t)x_j(t - \tau_q)\right) = \begin{cases} \rho_i r_{ij}(0) & \text{if} \quad i + j = 2q \\ 0 & \text{if} \quad i + j \neq 2q. \end{cases}$$ (4.13)

Lemmas 4.1 and 4.2 show the properties of the coded signals $x_1(t), x_2(t), \ldots, x_r(t)$. These properties can be utilized to separate the coded signals from their mixtures and then to estimate the source signals $s_1(t), s_2(t), \ldots, s_r(t)$.

Source Separation

Let $\mathbf{B} \in \mathbb{R}^{r \times m}$ be a scalar matrix and

$$\mathbf{u}(t) = \mathbf{B}\mathbf{y}(t) = \mathbf{C}\mathbf{x}(t)$$ (4.14)

where $\mathbf{C} = \mathbf{B}\mathbf{A} \in \mathbb{R}^{r \times r}$. Denote the ith row vector of $\mathbf{B}$ by $\mathbf{b}_i^T$, the ith row vector of $\mathbf{C}$ by $\mathbf{c}_i^T$, the ith element of $\mathbf{u}(t)$ by $u_i(t)$, and the (i, j)th entry of $\mathbf{C}$ by c_{ij}. Then, from (4.14), it follows

$$u_i(t) = \mathbf{b}_i^T \mathbf{y}(t) = \mathbf{c}_i^T \mathbf{x}(t)$$ (4.15)

for $i = 1, 2, \ldots, r$. To ensure that $u_i(t)$ is an estimate of $x_i(t)$, we need to find a vector $\mathbf{b}_i^T$ such that all elements of $\mathbf{c}_i^T$, except for the ith element c_{ii}, are zero. The separation criterion is shown in the following theorem [4]:

Theorem 4.1 *For each* $i \in \{1, 2, \ldots, r\}$, $\mathbf{b}_i^T$ *is a separation vector that extracts* $x_i(t)$ *(up to a scalar) if and only if*

$$E\left(u_i(t)u_i(t-\tau_q)\right) = 0 \tag{4.16}$$

$$E\left(u_i(t)u_i(t)\right) \neq 0 \tag{4.17}$$

for all $q \in \{1, 2, \ldots, r\} \setminus \{i\}$.

Now, we present an algorithm to separate the coded signals $\mathbf{x}(t)$ from $\mathbf{y}(t)$ by ensuring (4.16) and (4.17). By substituting $u_i(t) = \mathbf{b}_i^T \mathbf{y}(t)$ into (4.16) and (4.17), it yields

$$\mathbf{b}_i^T \mathbf{R}_{\mathbf{yy}}(\tau_q)\mathbf{b}_i = 0 \quad \text{for all} \ q \neq i \tag{4.18}$$

$$\mathbf{b}_i^T \mathbf{R}_{\mathbf{yy}}(0)\mathbf{b}_i \neq 0 \tag{4.19}$$

where $\mathbf{R}_{\mathbf{yy}}(\tau) = E\left(\mathbf{y}(t)\mathbf{y}(t-\tau)^T\right)$ and $i = 1, 2, \ldots, r$. One way to solve this problem is to minimize a cost function derived from (4.18) and (4.19). The cost function is usually of higher-order as the matrix $\mathbf{R}_{\mathbf{yy}}(\tau_q)$ is in general not positive definite. As a result, the algorithm derived from it is often not globally convergent and/or has low convergence rate. In order to overcome this problem, one can solve the equations in (4.18) sequentially in the following order:

$$\begin{cases} \mathbf{b}_i^T \mathbf{R}_{\mathbf{yy}}(\tau_{q_{r-1}})\mathbf{b}_i = 0 \\ \mathbf{b}_i^T \mathbf{R}_{\mathbf{yy}}(\tau_{q_{r-2}})\mathbf{b}_i = 0 \\ \quad \vdots \\ \mathbf{b}_i^T \mathbf{R}_{\mathbf{yy}}(\tau_{q_1})\mathbf{b}_i = 0 \end{cases} \tag{4.20}$$

where $q_i, i = 1, 2, \ldots, r$ are defined below

$$\begin{cases} \left[q_{r-1}, q_{r-2}, \ldots, q_1\right] = [r, r-1, \ldots, 2], & \text{if } i = 1 \\ \left[q_{r-1}, q_{r-2}, \ldots, q_1\right] = [1, 2, \ldots, r-1], & \text{if } i = r \\ \left[q_{r-1}, \ldots, q_i, q_{i-1}, \ldots, q_1\right] = [r, \ldots, i+1, 1, \ldots, i-1], & \text{if } 2 \leq i \leq r-1 \end{cases} \tag{4.21}$$

Obviously, the vector $\mathbf{b}_i$ satisfying (4.19) and (4.20) is a separation vector that extracts the ith coded signal $x_i(t)$. From the estimates of the separation vectors $\mathbf{b}_1^T, \mathbf{b}_2^T, \ldots, \mathbf{b}_r^T$, the coded signals can be obtained by

$$\hat{x}_i(t) = \mathbf{b}_i^T \mathbf{y}(t) \tag{4.22}$$

for $i = 1, 2, \ldots, r$.

Moreover, from (4.10), it follows

$$\hat{s}_i(t - l_i) = \hat{x}_i(t) - \rho_i \hat{s}_i(t - l_i') \tag{4.23}$$

Table 4.1 Precoding based time-domain algorithm [4]

Step 1	Estimate the matrices $\mathbf{R}_{yy}(0)$ and $\mathbf{R}_{yy}(\tau_q)$, $q = 1, 2, \ldots, r$
Step 2	Compute $\mathbf{b}_i$, $i = 1, 2, \ldots, r$
	(i) For each i, find the corresponding integers $q_1, q_2, \ldots, q_{r-1}$ from (4.21)
	(ii) Compute $\mathbf{P}_{r-1}$, which is composed of the $m - 1$ left singular vectors corresponding to the zero singular value of $\mathbf{R}_{yy}(\tau_{q_{r-1}})$
	(iii) Recursively compute matrices $\mathbf{P}_{r-k}$, $k = 2, 3, \ldots, r - 1$ by $\mathbf{P}_{r-k} = \mathbf{P}_{r-k+1}\mathbf{U}_{r-k+1}$, where $\mathbf{U}_{r-k+1}$ is a matrix consisting of the left singular vectors corresponding to the $m - k$ smallest singular values of $\mathbf{P}_{r-k+1}^T \mathbf{R}_{yy}(\tau_{q_{r-k}})\mathbf{P}_{r-k+1}$
	(iv) $\mathbf{b}_i$ is given by any column vector of $\mathbf{P}_1$ satisfying $\mathbf{b}_i^T \mathbf{R}_{yy}(0)\mathbf{b}_i \neq 0$
Step 3	Estimate the coded signals $x_1(t), x_2(t), \ldots, x_r(t)$ from (4.22)
Step 4	Recover the source signals $s_1(t), s_2(t), \ldots, s_r(t)$ from (4.24)

where $i = 1, 2, \ldots, r$. Based on (4.8), (4.9) and (4.23), it holds that

$$
\begin{aligned}
\hat{s}_i(t) &= \hat{x}_i(t + l_i) - \rho_i \hat{s}_i(t + l_i - l_i') \\
&= \hat{x}_i(t + 2(r - i)K) - \rho_i \hat{s}_i(t - (4i - 1)K)
\end{aligned}
\tag{4.24}
$$

where $i = 1, 2, \ldots, r$. Then the source signals can be recovered by using (4.24). The proposed algorithm is summarized in Table 4.1.

We would like to note that in order to recover $\mathbf{s}(t)$ from $\mathbf{x}(t)$, each precoder must be reversible by a stable filter. This requires $|\rho_i| < 1$. On the other hand, if the value of $|\rho_i|$ is too small, the correlation matrix $\mathbf{R}_{xx}(\tau_q)$ tends to be ill-conditioned [see (4.12) and (4.13)], which will have negative impact on the performance of source separation.

Simulation Results

We consider the case of $m = 4$ and $r = 3$. The three mutually correlated source signals are generated by

$$s_i(t) = \alpha_i(z)\varepsilon(t), \quad i = 1, 2, 3$$

where $\varepsilon(t)$ is a temporally white sequence randomly chosen from a uniform distribution on the interval $(-0.5, \ 0.5)$ and $\alpha_i(z)$, $i = 1, 2, 3$ are chosen as: $\alpha_1(z) = 1 - 0.43z^{-1} + 0.28z^{-2} + 1.19z^{-3}$, $\alpha_2(z) = 1 + 0.13z^{-1} + 1.75z^{-2} - 0.33z^{-3}$ and $\alpha_3(z) = 1 - 1.2z^{-1} - z^{-2} + 1.18z^{-3}$. For these source signals, the length of color is $M - 1 = 3$ (i.e., $M = 4$). In addition, we choose $\rho_1 = \rho_2 = \rho_3 = 0.8$ and the channel matrix $\mathbf{A}$ is randomly generated in each simulation run.

Under these simulation parameters, the performance of the algorithm in [4] is examined. The result of coded signal separation is evaluated using the mean interference rejection level (MIRL):

$$
\mathrm{MIRL(dB)} = 10 \log_{10} \left(\frac{1}{r(r-1)} \sum_{i=1}^{r} \sum_{\substack{j=1 \\ j \neq i}}^{r} \frac{E\left|(\mathbf{BA})_{ij}\right|^2}{E\left|(\mathbf{BA})_{ii}\right|^2} \right)
\tag{4.25}
$$

and the result of source signal estimation is assessed using the normalized mean-squared-error (NMSE) [5]:

$$\text{NMSE(dB)} = 10\log_{10}\left(\frac{\min_{\mathbf{D}}\sum_{t=0}^{n-1}\left\|\mathbf{D}\hat{\mathbf{s}}(t) - \mathbf{s}(t)\right\|^2}{\sum_{t=0}^{n-1}\|\mathbf{s}(t)\|^2}\right) \tag{4.26}$$

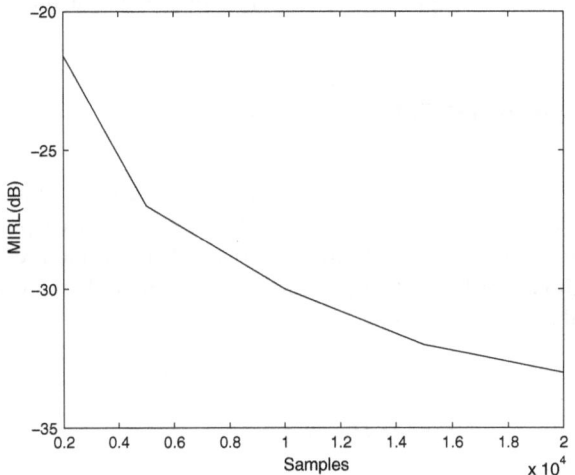

Fig. 4.2 Performance of separating the coded signals by the algorithm in [4], where $K = 4$

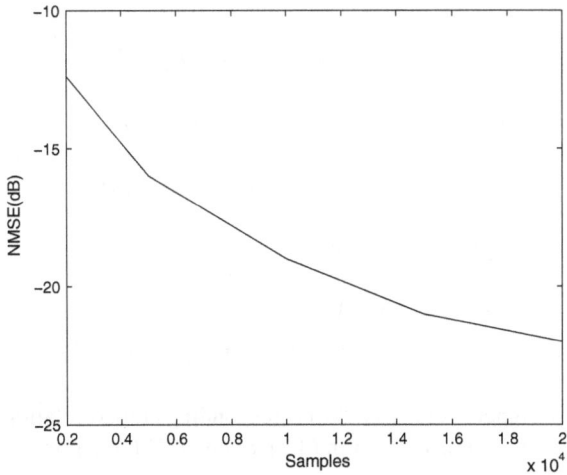

Fig. 4.3 Performance of recovering the source signals by the algorithm in [4], where $K = 4$

where $\hat{\mathbf{s}}(t) = \left[\hat{s}_1(t), \hat{s}_2(t), \ldots, \hat{s}_r(t)\right]^T$ and $\mathbf{D}$ is an $r \times r$ diagonal matrix. Clearly, the smaller MIRL and NMSE values, the better performance. These performance indices are calculated by averaging the results of 100 independent runs.

Figure 4.2 shows the performance of the algorithm in [4] in separating the coded signals. We can see that the algorithm yields small MIRL when the sample size is moderate or large. Moreover, increasing sample size leads to better separation result. The performance of the algorithm in recovering the source signals is shown in Fig. 4.3. As expected, the result of source recovery is satisfactory.

4.3 Precoding Based Z-Domain Methods

In this section, we present two precoding based Z-domain methods for separating mutually correlated sources. The first method uses second-order precoders, whilst the second one employs only first-order precoders. Clearly, the precoder orders of these two methods are lower than the order of their time-domain counterpart [4] introduced in the previous section.

4.3.1 Using Second-Order Precoders

Consider the system in Fig. 4.1. Assume that the precoders are second-order and denote the ith precoder by

$$
\begin{aligned}
p_i(z) &= 1 + p_{i,1}z^{-l} + p_{i,2}z^{-2} \\
&= (1 - r_{i,1}z^{-1})(1 - r_{i,2}z^{-1})
\end{aligned}
\tag{4.27}
$$

where $r_{i,1}$ and $r_{i,2}$ are the zeros of the precoder $p_i(z)$ and

$$
p_{i,1} = -(r_{i,1} + r_{i,2}) \quad \text{and} \quad p_{i,2} = r_{i,1}r_{i,2}.
$$

Then, the ith coded signal $x_i(t)$ can be written as

$$
x_i(t) = p_i(z)s_i(t)
\tag{4.28}
$$

$$
= s_i(t) + p_{i,1}s_i(t-1) + p_{i,2}s_i(t-2).
\tag{4.29}
$$

For the source signal vector $\mathbf{s}(t)$, the corresponding autocorrelation matrix at time lag τ, denoted as $\mathbf{R}_{ss}(\tau)$, is defined in (4.3). Same as [5], the power spectral matrix of $\mathbf{s}(t)$ is defined as

$$
\mathbf{Q}_{ss}(z) = \sum_{t=-\infty}^{\infty} \mathbf{R}_{ss}(t)z^{-t}.
\tag{4.30}
$$

Furthermore, let

$$\bar{\mathbf{s}}_i = [s_i(1), s_i(2), \ldots, s_i(t)]^T$$

where n is the number of samples of the source signals and $i = 1, 2, \ldots, r$.

In order to achieve the task of blindly separating mutually correlated sources using second-order precoders, the assumption (4.1) about the mixing matrix $\mathbf{A}$ (which is shown in Sect. 4.2) is needed, together with the following three assumptions:

(A4.4) The source signals $s_1(t), s_2(t), \ldots, s_r(t)$ are zero-mean, stationary, temporally white and mutually correlated but satisfy

$$[\bar{\mathbf{s}}_1, \bar{\mathbf{s}}_2, \ldots, \bar{\mathbf{s}}_r] \cdot \mathbf{d} \neq \mathbf{0} \tag{4.31}$$

for any nonzero vector $\mathbf{d} = [d_1, d_2, \ldots, d_r]^T$.

(A4.5) The noise signals $w_1(t), w_2(t), \ldots, w_m(t)$ are white Gaussian noise signals independent of the source signals.

(A4.6) All zeros of the precoders are distinct and satisfy $0 < |r_{i,l}| < 1$, where $r_{i,1} + r_{i,2}$ is a nonzero real number, $i = 1, 2, \ldots, r$ and $l = 1, 2$. They are known at the receiver.

Both Assumptions (A4.1) and (A4.5) have been used in various BSS approaches. Assumption (A4.6) presents a requirement on the zeros of the precoders. Equation (4.31) in Assumption (A4.4) means that any source signal $s_i(t)(i = 1, 2, \ldots, r)$ is not a linear combination of the other source signals. This assumption also implies that $E\left(s_i(t)s_j(t - \tau)\right) \neq 0$ with $i \neq j$ for some time lag(s) τ. Considering that the number of samples of the source signals is limited, we can suppose that there is a constant positive integer $\mu > 0$, such that

$$\beta_{i,j}(\tau) = E\left(s_i(t)s_j(t - \tau)\right) = 0, \ \forall \tau \notin [-\mu \ \mu] \tag{4.32}$$

for any i, j.

From the above equation and Assumption (A4.2), the following lemma is proposed [6].

Lemma 4.3 *The source power spectral matrix $\mathbf{Q}_{ss}(z)$ is singular for at most 2μ different z.*

Let us assume that the set $\tilde{\mathscr{Z}}$ consists of all complex numbers z_q making $\mathbf{Q}_{ss}(z_q)$ be singular. In other words, $\mathbf{Q}_{ss}(z)$ is of full rank for any $z \notin \tilde{\mathscr{Z}}$. Lemma 4.3 means that the number of elements in the set $\tilde{\mathscr{Z}}$ does not exceed 2μ. Since the zeros $r_{i,l}$ ($i = 1, 2, \ldots, r$ and $l = 1, 2$) of the precoders are randomly selected as any values satisfying Assumption (A4.6), they are not included in the finite set $\tilde{\mathscr{Z}}$, i.e., $r_{i,l} \notin \tilde{\mathscr{Z}}$, with probability one. This observation leads to the conclusion that the source power spectral matrix $\mathbf{Q}_{ss}(z)$ is of full rank at $z = r_{i,l}$ with probability one, where $i = 1, 2, \ldots, r$ and $l = 1, 2$. Next, we show how to exploit the Z-domain properties of the precoders given in (4.27) to find the separation matrix $\mathbf{B}$, and then to retrieve the source signals.

Separation Criterion

Similar to (4.3) and (4.30), we denote the autocorrelation matrices and power spectral matrices of $\mathbf{w}(t)$ and $\mathbf{y}(t)$ by $\mathbf{R}_{ww}(\tau)$, $\mathbf{R}_{yy}(\tau)$, $\mathbf{Q}_{ww}(z)$ and $\mathbf{Q}_{yy}(z)$, respectively. Since $\mathbf{y}(t) = \mathbf{A}\mathbf{x}(t) + \mathbf{w}(t)$, it holds from Assumption (A4.5) that

$$\mathbf{R}_{yy}(\tau) = \mathbf{A}\mathbf{R}_{xx}(\tau)\mathbf{A}^T + \mathbf{R}_{ww}(\tau) \tag{4.33}$$

where $\mathbf{R}_{ww}(\tau) \neq \mathbf{0}$ if $k = 0$ and $\mathbf{R}_{ww}(\tau) = \mathbf{0}$ if $k \neq 0$. From (4.28) and (4.33), it follows that

$$\begin{aligned}
\mathbf{Q}_{yy}(z) &= \sum_{\tau=-\infty}^{\infty} \mathbf{R}_{yy}(\tau)z^{-\tau} \\
&= \mathbf{A}\mathbf{Q}_{xx}(z)\mathbf{A}^T + \mathbf{Q}_{ww}(z) \\
&= \mathbf{A}\mathbf{P}(z)\mathbf{Q}_{ss}(z)\mathbf{P}((z^*)^{-1})^H\mathbf{A}^T + \mathbf{C}_{ww}(0)
\end{aligned} \tag{4.34}$$

where $\mathbf{P}(z) = diag(p_1(z), p_2(z), \ldots, p_r(z))$, and the superscripts $*$ and ^{H} stand for complex conjugate and complex conjugate transpose, respectively.

Let $\mathbf{P}_i(z)$ be equal to $\mathbf{P}(z)$ with the ith diagonal entry replaced by zero, i.e.,

$$\mathbf{P}_i(z) = diag(p_1(z), \ldots, p_{i-1}(z), 0, p_{i+1}(z), \ldots, p_N(z)) \tag{4.35}$$

and

$$\mathbf{T}_i(z) = \mathbf{P}_i(z)\mathbf{Q}_{ss}(z)\mathbf{P}((z^*)^{-1})^H \tag{4.36}$$

where $i = 1, 2, \ldots, r$. The matrix $\mathbf{T}_i(r_{i,1})$ has the following property [6].

Lemma 4.4 *For any $i \in \{1, 2, \ldots, r\}$, all rows in the matrix $\mathbf{T}_i(r_{i,1})$ excluding the ith row must be linearly independent.*

Similarly, the ith row of $\mathbf{T}_i(r_{i,2})$ is also all-zero, and all other rows of $\mathbf{T}_i(r_{i,2})$ are linearly independent. Then, the ith row of the matrix $\mathbf{T}_i(r_{i,1}) - \mathbf{T}_i(r_{i,2})$ must also be a zero vector. Since $r_{i,1} \neq r_{i,2}$, Lemma 4.4 results in the conclusion that for any $i \in \{1, 2, \ldots, r\}$, all rows in the matrix $\mathbf{T}_i(r_{i,1}) - \mathbf{T}_i(r_{i,2})$ excluding the i-th row are linearly independent with probability one. This implies that the rank of the matrix $\mathbf{T}_i(r_{i,1}) - \mathbf{T}_i(r_{i,2})$ is $r - 1$. This conclusion lays the theoretical foundation for the proposition of the separation criterion.

To proceed, let $\mathbf{A}_i$ equal $\mathbf{A}$ with its ith column replaced by a zero vector, i.e.,

$$\mathbf{A}_i = [\mathbf{a}_1, \ldots, \mathbf{a}_{i-1}, \mathbf{0}, \mathbf{a}_{i+1}, \ldots, \mathbf{a}_N]. \tag{4.37}$$

From (4.34), it holds that

$$
\begin{aligned}
\mathbf{Q}_{\mathbf{yy}}(r_{i,1}) &= \mathbf{A}\mathbf{P}(r_{i,1})\mathbf{Q}_{\mathbf{ss}}(r_{i,1})\mathbf{P}((r_{i,1}^*)^{-1})^H \mathbf{A}^T + \mathbf{C}_{\mathbf{ww}}(0) \\
&= \mathbf{A}_i \mathbf{P}_i(r_{i,1})\mathbf{Q}_{\mathbf{ss}}(r_{i,1})\mathbf{P}((r_{i,1}^*)^{-1})^H \mathbf{A}^T + \mathbf{C}_{\mathbf{ww}}(0)
\end{aligned} \tag{4.38}
$$

and

$$
\mathbf{Q}_{\mathbf{yy}}(r_{i,2}) = \mathbf{A}_i \mathbf{P}_i(r_{i,2})\mathbf{Q}_{\mathbf{ss}}(r_{i,2})\mathbf{P}((r_{i,2}^*)^{-1})^H \mathbf{A}^T + \mathbf{C}_{\mathbf{ww}}(0). \tag{4.39}
$$

Let

$$
\mathbf{Q}_i = \mathbf{Q}_{\mathbf{yy}}(r_{i,1}) - \mathbf{Q}_{\mathbf{yy}}(r_{i,2}). \tag{4.40}
$$

Based on (4.38)–(4.40), it follows

$$
\begin{aligned}
\mathbf{Q}_i &= \mathbf{A}_i \left(\mathbf{P}_i(r_{i,1})\mathbf{Q}_{\mathbf{ss}}(r_{i,1})\mathbf{P}((r_{i,1}^*)^{-1})^H \right. \\
&\qquad \left. -\mathbf{P}_i(r_{i,2})\mathbf{Q}_{\mathbf{ss}}(r_{i,2})\mathbf{P}((r_{i,2}^*)^{-1})^H \right) \mathbf{A}^T
\end{aligned} \tag{4.41}
$$

$$
= \mathbf{A}_i (\mathbf{T}_i(r_{i,1}) - \mathbf{T}_i(r_{i,2}))\mathbf{A}^T. \tag{4.42}
$$

Then, the separation criterion is proposed in the theorem below [6].

Theorem 4.2 $\mathbf{b}_i^H$ *is a separation vector ensuring*

$$
\mathbf{b}_i^H \mathbf{A} = [0, \ldots, 0, c_i, 0, \ldots, 0] \tag{4.43}
$$

if and only if

$$
\begin{cases} \mathbf{b}_i^H \mathbf{Q}_i = 0 \\ \mathbf{b}_i^H \mathbf{R}_{\mathbf{yy}}(\tau)\mathbf{b}_i \neq 0 \end{cases} \tag{4.44}
$$

where $c_i \neq 0$, $\tau = 1, 2$, and $i = 1, 2, \ldots, r$.

It is also shown in [6] that the condition (4.44) with the time lags $\tau = 0$ and $\tau \geq 3$ cannot guarantee that the obtained $\mathbf{b}_i^H$ is a separation vector. In other words, τ can only take value 1 or 2. Without loss of generality, we choose $\tau = 1$.

Source Separation
Next, an algorithm will be introduced to perform BSS, which is composed of two procedures, i.e., estimation of the separation vectors and recovery of the source signals. Firstly, let us discuss how to estimate the separation vectors. As mentioned above, the rank of the matrix $\mathbf{T}_i(r_{i,1}) - \mathbf{T}_i(r_{i,2})$ is $r - 1$. Thus, from (4.42), one can infer that the rank of the matrix $\mathbf{Q}_i$ is also $r - 1$. Since $\mathbf{Q}_i$ is an $m \times m$ matrix, it has $m - r + 1$ singular values which are zero. Also, since $m \geq r$, the matrix $\mathbf{Q}_i$ should have at least one zero singular value.

Let $\mathbf{V}_i$ be an $m \times (m-r+1)$ matrix, whose columns are formed by the $m-r+1$ left singular vectors corresponding to the zero singular value of $\mathbf{Q}_i$.[1] Assume that the column vector $\mathbf{u}$ is the eigenvector corresponding to any nonzero eigenvalue $\tilde{\lambda}$ of the matrix $\mathbf{V}_i^H \mathbf{R}_{\mathbf{yy}}(1) \mathbf{V}_i$. One can verify that

$$\mathbf{u}^H \mathbf{V}_i^H \mathbf{Q}_i = 0$$

and

$$\mathbf{u}^H \mathbf{V}_i^H \mathbf{R}_{\mathbf{yy}}(1) \mathbf{V}_i \mathbf{u} = \tilde{\lambda} \cdot \|\mathbf{u}\|^2 \neq 0.$$

Then, the separation vector $\mathbf{b}_i$ can be selected as $\mathbf{b}_i = \mathbf{V}_i \mathbf{u}$, which satisfies (4.44) with $\tau = 1$.

After the separation vectors $\mathbf{b}_1^H, \mathbf{b}_2^H, \ldots, \mathbf{b}_r^H$ are obtained, the coded signals can be estimated as follows:

$$\hat{x}_i(t) = \mathbf{b}_i^H \mathbf{y}(t) \tag{4.45}$$

for $i = 1, 2, \ldots, r$. Based on the estimates of the coded signals, the source signals can be obtained using the following equation:

$$\hat{s}_i(t) = \hat{x}_i(t) - p_{i,1} \hat{s}_i(t-1) - p_{i,2} \hat{s}_i(t-2) \tag{4.46}$$

where $i = 1, 2, \ldots, r$, which is derived from (4.29). The complete algorithm is summarized in Table 4.2.

Simulation Results

The mixing system considered has three sources ($r = 3$) and four mixtures ($m = 4$). The zeros of the precoders are chosen as

$$r_{i,1} = \eta_i e^{j\pi\left(\frac{2i-1}{6}\right)} \quad \text{and} \quad r_{i,2} = r_{i,1}^* \tag{4.47}$$

where $j = \sqrt{-1}$, $\eta_i = 0.5$, and $i = 1, 2, 3$. Clearly, the precoder zeros selected according to (4.47) satisfy Assumption (A4.6) and are equally spaced in angle in the Z-plane. Let $\mathbf{E}(t) = [\varepsilon_1(t), \varepsilon_2(t), \varepsilon_3(t)]^T$, where $\varepsilon_i(t)$, $i = 1, 2, 3$ are temporally white sequences randomly chosen from a normal distribution $(0, 1)$. Set

$$\mathbf{M} = \begin{bmatrix} 1 & 0.7012 & 0.5649 \\ 0.7012 & 1 & 0.6123 \\ 0.5649 & 0.6123 & 1 \end{bmatrix}$$

---

[1] In practical applications, those singular values that should be zero in theory could become small but not necessarily zero, due to the presence of noise and small errors during the estimation of the matrix $\mathbf{Q}_i$. An approximate solution is to find the left singular vectors associated with the $m-r+1$ smallest singular values of $\mathbf{Q}_i$.

Table 4.2 Precoding based Z-domain algorithm using second-order precoders [6]

Step 1	Compute $\mathbf{R_{yy}}(\tau) \approx \frac{1}{n}\sum_{t=0}^{n-1}\mathbf{y}(t)\mathbf{y}(t-\tau)^T$ where n is the number of samples of the mixtures
Step 2	Compute $\mathbf{Q_{yy}}(r_{i,1})$ and $\mathbf{Q_{yy}}(r_{i,2})$ by $\mathbf{Q_{yy}}(z) \approx \sum_\tau \mathbf{R_{yy}}(\tau)z^{-\tau}$ where $i = 1, 2, \ldots, r$
Step 3	Obtain $\mathbf{Q}_i$ from (4.40), where $i = 1, 2, \ldots, r$
Step 4	For any given $i(i = 1, 2, \ldots, r)$, find the left singular vectors $\mathbf{v}_{i,1}, \mathbf{v}_{i,2}, \ldots, \mathbf{v}_{i,m-r+1}$ corresponding to the $m - r + 1$ smallest singular values of $\mathbf{Q}_i$, and construct the matrix $\mathbf{V}_i = [\mathbf{v}_{i,1}, \mathbf{v}_{i,2}, \ldots, \mathbf{v}_{i,m-r+1}]$
Step 5	Compute all the eigenvalues of the matrix $\mathbf{V}_i^H \mathbf{R_{yy}}(1)\mathbf{V}_i$ as $\lambda_1, \lambda_2, \ldots, \lambda_{m-r+1}$ ordered by $|\lambda_1| \geq |\lambda_2| \geq \cdots \geq |\lambda_{m-r+1}|$, and their corresponding eigenvectors $\mathbf{u}_1, \mathbf{u}_2, \ldots, \mathbf{u}_{m-r+1}$
Step 6	The separation vector $\mathbf{b}_i$ can be obtained as $\mathbf{b}_i = \mathbf{V}_i\mathbf{u}_1$
Step 7	Estimate the coded signals $x_1(t), x_2(t), \ldots, x_r(t)$ from (4.45)
Step 8	Estimate the source signals $s_1(t), s_2(t), \ldots, s_r(t)$ from (4.46)

and denote its eigenvalue decomposition by $[\mathbf{U}, \mathbf{S}] = \mathrm{eig}(\mathbf{M})$. Then the mutually correlated source signals are generated by

$$\mathbf{s}(t) = [s_1(t), s_2(t), s_3(t)]^T = \left(\mathbf{U}\mathbf{S}^{0.5}\mathbf{U}^H\right)\mathbf{E}(t).$$

Since the off-diagonal entries of $\mathbf{M}$ are much greater than zero, the source signals are highly correlated. Furthermore, the channel matrix $\mathbf{A}$ is randomly generated in each simulation run and the signal to noise ratio (SNR) is defined as SNR$= -10\log_{10}\sigma_w^2$, where σ_w^2 is the variance of the noise signals.

We evaluate the effectiveness of the second-order Z-domain algorithm presented in this subsection. The performance of source signal estimation is measured by the NMSE index defined in (4.26). The NMSE index is estimated by averaging 100 independent runs, where SNR $= 30$ dB and $n = 15, 000$. As shown in Fig. 4.4, the algorithm achieves small NMSE when the sample size is reasonably large.

4.3.2 Using First-Order Precoders

In the previous subsection, the Z-domain features of the precoders are utilized, where the order of the precoders is set to two, regardless of the number of the sources [6]. In this subsection, it will be shown that the order of the precoders can be further reduced to one. The usage of first-order precoders shortens the delay in data transmission and simplifies the implementation of the precoding scheme.

Consider the system in Fig. 4.1, which lead to the corresponding system model $\mathbf{y}(t) = \mathbf{A}\mathbf{x}(t) + \mathbf{w}(t)$. Denote the ith precoder by

$$p_i(z) = 1 - r_i z^{-1} \tag{4.48}$$

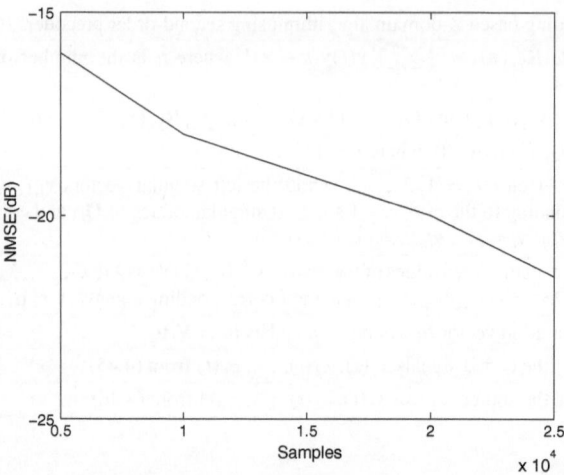

Fig. 4.4 Performance of recovering the source signals by the second-order Z-domain algorithm in [6]

where r_i is the zero of the precoder $p_i(z)$ and $i = 1, 2, \ldots, r$. Thus, the ith coded signal can be written as

$$x_i(t) = p_i(z)s_i(t) \tag{4.49}$$
$$= s_i(t) - r_i s_i(t-1) \tag{4.50}$$

where $i = 1, 2, \ldots, r$. Given $\mathbf{s}(t) = [s_1(t), s_2(t), \ldots, s_r(t)]^T$, one can compute $\mathbf{R}_{ss}(\tau)$ from (4.3) and $\mathbf{Q}_{ss}(z)$ from (4.30), which are the autocorrelation matrix of $\mathbf{s}(t)$ and the power spectral matrix of $\mathbf{s}(t)$, respectively.

Assume

(A4.7) The system has more outputs than inputs (i.e., $m > r \geq 2$), and the channel matrix $\mathbf{A}$ is of full column rank.

(A4.8) The source signals $s_1(t), s_2(t), \ldots, s_r(t)$ are zero-mean, stationary and temporally white but spatially correlated. The source power spectral matrix $\mathbf{Q}_{ss}(z)$ is of full rank at $z = r_i$, where $i = 1, 2, \ldots, r$.

(A4.9) The noises $w_1(t), w_2(t), \ldots, w_m(t)$ are temporally white and mutually uncorrelated with zero mean and equal variance σ_w^2. They are also independent of the source signals.

(A4.10) The zeros of the precoders, r_i, $i = 1, 2, \ldots, r$, are distinct and satisfy $|r_i| < 1$. All precoder zeros are known at the receiver.

We would like to note that since the selection of the precoder zeros is irrelevant to the source signals, the source power spectral matrix $\mathbf{Q}_{ss}(z)$ has full rank at $z = r_i$ with probability one.

Similar to the definitions of $\mathbf{R}_{ss}(\tau)$ and $\mathbf{Q}_{ss}(z)$, one can define the autocorrelation matrices and power spectral matrices of $\mathbf{x}(t)$, $\mathbf{w}(t)$ and $\mathbf{y}(t)$, and denote them as

$\mathbf{R_{xx}}(\tau)$, $\mathbf{R_{ww}}(\tau)$, $\mathbf{R_{yy}}(\tau)$, $\mathbf{Q_{xx}}(z)$, $\mathbf{Q_{ww}}(z)$ and $\mathbf{Q_{yy}}(z)$, respectively. It is shown in (4.33) that $\mathbf{R_{yy}}(\tau) = \mathbf{A}\mathbf{R_{xx}}(\tau)\mathbf{A}^T + \mathbf{R_{ww}}(\tau)$. From Assumption (A4.9), $\mathbf{R_{ww}}(\tau) \neq \mathbf{0}$ if $\tau = 0$ and $\mathbf{R_{ww}}(\tau) = \mathbf{0}$ if $\tau \neq 0$. Hence, it holds that

$$\mathbf{R_{yy}}(0) = \mathbf{A}\mathbf{R_{xx}}(0)\mathbf{A}^T + \sigma_w^2\mathbf{I}_m \tag{4.51}$$

where $\mathbf{I}_m$ denotes the $m \times m$ identity matrix. Since $\mathbf{A}$ is a tall matrix, (4.51) implies that σ_w^2 is the smallest eigenvalue of $\mathbf{R_{yy}}(0)$ and thus can be estimated from $\mathbf{R_{yy}}(0)$. Then, $\sigma_w^2\mathbf{I}_m$ can be easily removed from $\mathbf{R_{yy}}(0)$, i.e.,

$$\mathbf{R_{yy}}(0) \longleftarrow \mathbf{R_{yy}}(0) - \sigma_w^2\mathbf{I}_m. \tag{4.52}$$

As a result, (4.34) reduces to

$$\mathbf{Q_{yy}}(z) = \mathbf{A}\mathbf{P}(z)\mathbf{Q_{ss}}(z)\mathbf{P}((z^*)^{-1})^H\mathbf{A}^T \tag{4.53}$$

where $\mathbf{P}(z) = diag(p_1(z), p_2(z), \ldots, p_r(z))$.
From (4.35)–(4.37) and (4.53), it yields

$$\begin{aligned}
\mathbf{Q_{yy}}(r_i) &= \mathbf{A}\mathbf{P}(r_i)\mathbf{Q_{ss}}(r_i)\mathbf{P}((r_i^*)^{-1})^H\mathbf{A}^T \\
&= \mathbf{A}_i\mathbf{P}_i(r_i)\mathbf{Q_{ss}}(r_i)\mathbf{P}((r_i^*)^{-1})^H\mathbf{A}^T \\
&= \mathbf{A}_i\mathbf{T}(r_i)\mathbf{A}^T.
\end{aligned} \tag{4.54}$$

It is shown in [6] that for any given $i \in \{1, 2, \ldots, r\}$, all rows of the matrix $\mathbf{T}(r_i)$ excluding the ith row are linearly independent. Based on this conclusion, the following separation criterion is proposed [7].

Theorem 4.3 $\mathbf{b}_i^H$ *is a separation vector ensuring* (4.43) *if and only if*

$$\begin{cases} \mathbf{b}_i^H\mathbf{Q_{yy}}(r_i) = \mathbf{0} \\ \mathbf{b}_i^H\mathbf{R_{yy}}(1)\mathbf{b}_i \neq 0 \end{cases} \tag{4.55}$$

where $i = 1, 2, \ldots, r$.

By exploiting the separation criterion proposed in Theorem 4.3, an algorithm can be developed to recover the mutually correlated source signals. Since all rows of the matrix $\mathbf{T}(r_i)$ excluding the ith row are linearly independent, the rank of $\mathbf{T}(r_i)$ must be $r-1$. From (4.54), the rank of $\mathbf{Q_{yy}}(r_i)$ is also $r-1$. Thus, $\mathbf{Q_{yy}}(r_i)$ possesses $m-r+1$ zero singular values. Obviously, any of the $m-r+1$ left singular vectors associated with the zero singular value of $\mathbf{Q_{yy}}(r_i)$ can be taken as $\mathbf{b}_i$ making $\mathbf{b}_i^H\mathbf{Q_{yy}}(r_i) = \mathbf{0}$. On the other hand, the chosen $\mathbf{b}_i$ should also satisfy $\mathbf{b}_i^H\mathbf{R_{yy}}(1)\mathbf{b}_i \neq 0$. Therefore, the obtained $\mathbf{b}_i$ satisfies both mathematical relations in (4.55). In practice, an approximate solution is to find the left singular vectors corresponding to the $m-r+1$ smallest singular values of $\mathbf{Q_{yy}}(r_i)$. After obtaining the estimates of the separation vectors

Table 4.3 Precoding based Z-domain algorithm using first-order precoders [7]

Step 1	Compute $\mathbf{R_{yy}}(\tau) \approx \dfrac{1}{n}\sum_{t=0}^{n-1}\mathbf{y}(t)\mathbf{y}(t-\tau)^T$, where n is the number of samples of the mixtures, and remove $\sigma_w^2\mathbf{I}_m$ from $\mathbf{R_{yy}}(0)$
Step 2	Compute $\mathbf{Q_{yy}}(r_i)$ by $\mathbf{Q_{yy}}(z) \approx \sum_\tau \mathbf{R_{yy}}(\tau)z^{-\tau}$, where $i = 1, 2, \ldots, r$
Step 3	For any given $i\,(1 \le i \le r)$, find the left singular vectors corresponding to the $m-r+1$ smallest singular values of $\mathbf{Q_{yy}}(r_i)$, and choose the one satisfying $\mathbf{b}_i^H\mathbf{R_{yy}}(1)\mathbf{b}_i \ne 0$ as the separation vector
Step 4	Estimate the coded signals by $\hat{x}_i(t) = \mathbf{b}_i^H\mathbf{y}(t)$, $i = 1, 2, \ldots, r$
Step 5	Estimate the source signals by $\hat{s}_i(t) = \hat{x}_i(t) + r_i\hat{s}_i(t-1)$, $i = 1, 2, \ldots, r$

$\mathbf{b}_1^H, \mathbf{b}_2^H, \ldots, \mathbf{b}_r^H$, the coded signals and source signals can be easily computed, as we have explained in the previous subsection. The algorithm is shown in Table 4.3.

Simulation Results

An instantaneous mixing system with $r = 3$ sources and $m = 4$ mixtures is considered in the simulations. Let $\varepsilon_1(t), \varepsilon_2(t)$ and $\varepsilon_3(t)$ be three temporally white sequences randomly chosen from a uniform distribution on the interval $(-0.5,\ 0.5)$. The three spatially correlated source signals are generated as follows:

(a) $s_1(t) = \varepsilon_1(t)$;
(b) $s_2(t)$ is composed of the odd-labelled samples of $\varepsilon_1(t)$ and the even-labelled samples of $\varepsilon_2(t)$; and
(c) $s_3(t)$ consists of the odd-labelled samples of $\varepsilon_3(t)$ and the even-labelled samples of $\varepsilon_1(t)$.

The zeros of the three first-order precoders used in the proposed algorithm are randomly chosen as $r_1 = -0.0952 - j0.3847$, $r_2 = 0.0380 + j0.2242$, and $r_3 = 0.3167 + j0.1499$, where $j = \sqrt{-1}$. The channel matrix $\mathbf{A}$ is randomly generated in each simulation run and the signal to noise ratio is defined as $\text{SNR} = -10\log_{10}\sigma_w^2$. The performance of coded signal separation is assessed by the MIRL index defined in (4.25) and the performance of source signal estimation is measured by the NMSE index defined in (4.26). These performance indices are estimated by averaging 50 independent runs.

The simulation results are shown in Figs. 4.5 and 4.6. One can see from Fig. 4.5 that the mean interference rejection level of the precoding based first-order Z-domain algorithm is satisfactory even at low SNRs, where the sample size is fixed at $n = 15, 000$. This result is expected. Since $\mathbf{Q_{yy}}(r_i)$ and $\mathbf{R_{yy}}(1)$ in (4.55) are independent of the additive noise signals, the presence of additive noise does not affect the estimation accuracy of $\mathbf{b}_i$ in theory and only has mild impact in practice. Furthermore, as shown in Fig. 4.6, the algorithm achieves small normalized mean-squared-error when the sample size is moderate or large. The more samples are used, the better source recovery result is obtained.

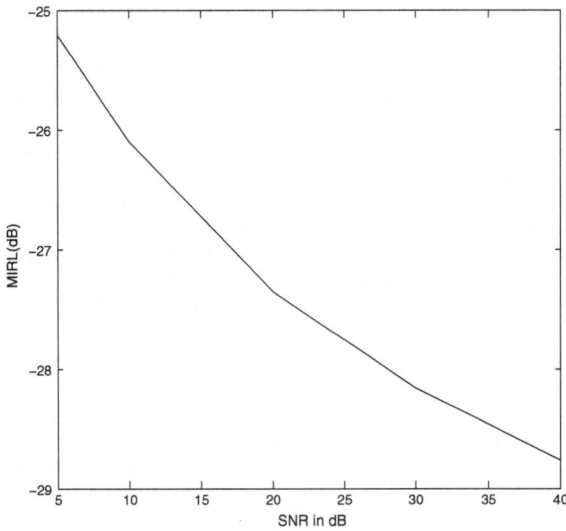

Fig. 4.5 Performance of separating the coded signals by the first-order Z-domain algorithm in [7], where the sample size $n = 15{,}000$

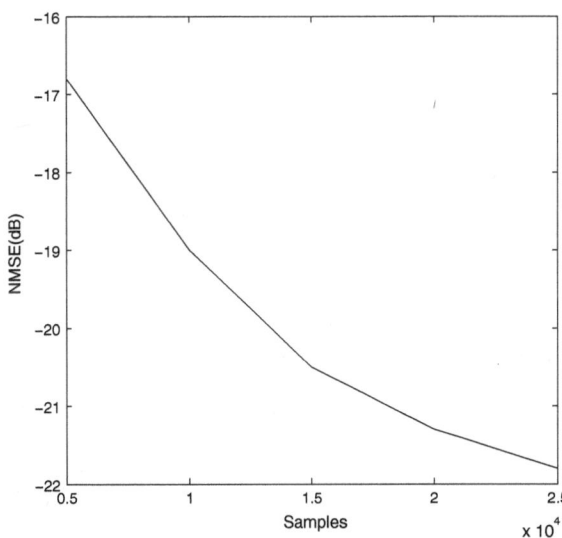

Fig. 4.6 Performance of recovering the source signals by the first-order Z-domain algorithm in [7], where SNR $= 30$ dB

References

1. W. Naanaa, J.-M. Nuzillard, Blind source separation of positive and partially correlated data. Sig. Proc. **85**(9), 1711–1722 (2005)
2. P.O. Hoyer, Non-negative matrix factorization with sparseness constraints. J. Mach. Learn. Res. **5**, 1457–1469 (2004)
3. D. Peng, Y. Xiang, Underdetermined blind separation of non-sparse sources using spatial time-frequency distributions. Digital Sig. Proc. **20**(2), 581–596 (2010)
4. Y. Xiang, S.K. Ng, V.K. Nguyen, Blind separation of mutually correlated sources using pre-coders. IEEE Trans. Neural Networks **21**(1), 82–90 (2010)
5. Y. Hua, S. An, Y. Xiang, Blind identification of FIR MIMO channels by decorrelating subchannels. IEEE Trans. Sig. Proc. **51**(5), 1143–1155 (2003)
6. Y. Xiang, D. Peng, Y. Xiang, S. Guo, Novel Z-domain precoding method for blind separation of spatially correlated signals. IEEE Trans. Neural Networks Learning Systems **24**(1), 94–105 (2013)
7. Y. Xiang, D. Peng, A. Kouzani, Separating spatially correlated signals using first-order precoders, in *Proceedings of The 9th IEEE Conference on Industrial Electronics and Applications*, China (2014)

Chapter 5
Future Work

Abstract In this chapter, we outline the technical issues and challenges in blind separation of mutually correlated sources. This points out directions for future work.

Keywords Nonnegativity · Sparsity · Time-frequency analysis · Precoding

Blind source separation (BSS) is a challenging problem in signal processing and it has potential applications in various real-world applications. While most BSS methods require the source signals to be independent or at least mutually uncorrelated, spatially correlated sources do exist in many applications. The mutual correlation among the source signals may result from the dense deployment of sensors [1], inherent correlation feature existing in some types of sources (such as images) [2–5], or multiple-input multiple-output wireless relay systems [6–9], to name some of them. The problem of blind separation of mutually correlated sources is an extremely difficult task as there is not sufficient statistical diversity among the sources.

To separate mutually correlated sources, one needs to exploit other properties of the source signals, such as nonnegativity and/or sparsity in time or time-frequency (TF) domain, or build some diversity among the source signals by using precoders. Based on these ideas, a number of methods were proposed to tackle the problem of blind separation of spatially correlated sources in the last few years. Different from conventional BSS techniques, these methods achieve BSS via dependent component analysis (DCA). However, the existing BSS methods via DCA are far from perfection. Much future work needs to be done to further improve these techniques and develop new blind separation techniques for mutually correlated sources.

5.1 Future Work for DCA Exploiting Nonnegativity and/or Time-Domain Sparsity

Nonnegativity is a useful feature for dependent source recovery and a series of nonnegativity based methods are introduced in Chap. 2, including methods based on nonnegativity sparse representation (NSR), convex geometry analysis (CGA) and nonnegative matrix factorization (NMF). For the methods based on NSR, they

© The Author(s) 2015   91
Y. Xiang et al., *Blind Source Separation*,
SpringerBriefs in Signal Processing, DOI 10.1007/978-981-287-227-2_5

jointly utilize the nonnegativity and sparsity features. One issue is how to employ the nonnegativity to construct advanced sparsity measure functions, which largely affect the cost functions of the NSR methods. Another issue is about the design of the corresponding algorithm. As we know, nonnegativity restricts the space of the solutions. Thus, more efficient algorithms could be developed under the nonnegativity constraint but how to design them is still an open problem.

For the CGA-based methods, they exploit the geometric feature along with the nonnegativity property, which brings a new viewpoint for DCA/BSS. Relatively speaking, the CGA-based methods have advantages in terms of the precision and speed of source recovery. However, they often work under strict conditions and these conditions may be violated by noise. How to improve their robustness against noise is a big challenge. In addition, the current geometric methods mainly focus on the case that the number of the sources is not greater than that of the mixtures. Is it possible to perform blind separation of spatially correlated sources in the underdetermined case by using CGA?

For the NMF-based methods, they exploit the nonnegativity of both the source signals and the mixing matrix. Since a NMF scheme does not necessarily generate the desired results, it needs to add other constraints to solve DCA/BSS. Due to the complexity of the real-world environment, developing proper constraints is a difficult task. Moreover, the convergence rate of the traditional NMF-based algorithms is slow and thus needs to be improved.

5.2 Future Work for DCA Exploiting Time-Frequency Analysis (TFA)

TFA is a powerful tool for DCA and considerable research efforts have been devoted to developing DCA approaches based on TFA. On the other hand, it is notable that the TFA-based DCA still faces two serious challenges. Firstly, let us briefly review the performance problem of TFA-based DCA. In Chap. 3, the quadratic TF distribution (TFD) is utilized to identify the mixing matrix and recover the source signals. Although the quadratic TFD can provide high TF resolution degree than the linear TFD, the former suffers from the so-called "cross terms" problem. The cross terms result from the energy distributions of two different source signals. The cross-term effect of the quadratic TFDs results in a "ghost" energy contribution, which significantly deteriorates the performance of TFA-based blind approaches. In future work, one could consider to exploit the reduced interference distribution to decrease or remove the negative effects of the cross terms of the quadratic TFDs. This could greatly improve the performance of TFA-based DCA methods.

Another challenge encountered by TFA-based DCA is its limited application scope, which is caused by several factors. (i) TFA is only suitable for nonstationary signals whose frequency contents vary with time. (ii) Although blind source recovery has been achieved successfully with non-sparse source signals, the estimation of the mixing matrix still needs to impose some sparsity constraints on the source signals. (iii) The performance of the TFA-based DCA can be easily affected by the TF "cross terms" among different source signals. Therefore, it is interesting and necessary to study how to relax the restrictions imposed on the source signals such that the TFA-based DCA approaches can be applied to a wider range of practical applications.

5.3 Future Work for DCA Exploiting Precoding

Building spatial diversity among the mutually correlated source signals by precoders is an interesting approach. The effectiveness of this approach has bee demonstrated by the three precoding based methods presented in Chap. 4. The advantage of the precoding based DCA is that it does not impose strong conditions on the source signals. For instance, it does not require the source signals to be nonnegative, sparse or locally dominant. However, the existing precoding based methods have some shortcomings that need to be overcome. For the precoding based time-domain method, the order of the precoders is proportional to the number of the sources. With the increase of the number of sources, the method becomes more complex and results in more transmission delay.

For the precoding based Z-domain methods, the performance of coded signal separation highly depends on the invariance of the precoder zeros. Strong additive noise and substantial calculation inaccuracy may change the values of precoder zeros and thus deteriorate the performance of the precoding based Z-domain methods. In addition, all of the precoding based methods (in both time-domain and Z-domain) need to recover the source signals from the estimates of the coded signals. This requires that each precoder must be reversible by a stable filter. Consequently, all zeros of the precoders must be within the unit circle in the Z-plane. Since the rise of the number of sources requires more precoders and thus more zeros in the unit circle, it shortens the distances among the zeros. This weakens the diversity of the coded signals, which unavoidably lowers the performance of codedd source separation.

Last but not least, all of the DCA methods shown in Chaps. 2–4 only consider instantaneous mixing systems. How to extend these methods to convolutional (or dynamic) mixing systems is another big challenge.

References

1. S.S. Pradhan, J. Kusuma, K. Ramchandran, Distributed compression in a dense microsensor network. IEEE Signal Process. Mag. **19**(2), 51–60 (2002)
2. C.-Y. Chang, A.A. Maciejewski, V. Balakrishnan, Fast eigenspace decomposition of correlated images. IEEE Trans. Image Process. **9**(11), 1937–1949 (2000)
3. K. Saitwal, A.A. Maciejewski, R.G. Roberts, B.A. Draper, Using the low-resolution properties of correlated images to improve the computational efficiency of eigenspace decomposition. IEEE Trans. Image Process. **15**(8), 2376–2387 (2006)

4. Y. Peng, A. Ganesh, J. Wright, W. Xu, Y. Ma, RASL: Robust alignment by sparse and low-rank decomposition for linearly correlated images, in *Proceedings 2010 IEEE Computer Society Conference on Computer Vision and Pattern Recognition*, pp. 763–770 (2000) art. no. 5540138
5. G. Zhou, Z. Yang, S. Xie, J. Yang, Online blind source separation using incremental nonnegative matrix factorization with volume constraint. IEEE Trans. Neural Netw. **22**(4), 550–560 (2011)
6. X. Tang, Y. Hua, Optimal design of non-regenerative MIMO wireless relays. IEEE Trans. Wireless Commun. **6**(4), 1398–1407 (2007)
7. I. Hammerström, A. Wittneben, Power allocation schemes for amplify-and-forward MIMO-OFDM relay links. IEEE Trans. Wireless Commun. **6**(8), 2798–2802 (2007)
8. A.S. Behbahani, R. Merched, A.M. Eltawil, Optimizations of a MIMO relay network. IEEE Trans. Signal Process. **56**(10), 5062–5073 (2008)
9. Y. Rong, M.R.A. Khandaker, Y. Xiang, Channel estimation of dual-hop MIMO relay system via parallel factor analysis. IEEE Trans. Wireless Commun. **11**(6), 2224–2233 (2012)